THE GENE AGE

THE
GENE AGE

Genetic Engineering
and the
Next Industrial Revolution

Edward J. Sylvester
and
Lynn C. Klotz

Charles Scribner's Sons · New York

Copyright © 1983 Edward J. Sylvester and Lynn C. Klotz

Library of Congress Cataloging in Publication Data

Sylvester, Edward J.
　The gene age.

　Includes bibliographical references and index.
　1. Genetic engineering.　I. Klotz, Lynn C.
II. Title.
TP248.6.S94　1983　　　575.1　　　83-11486
ISBN 0-684-17950-4

Printed in the United States of America.

*The photographs of human-growth-hormone-producing bacteria
appear courtesy of Genentech, Inc. The photograph of bacterial
DNA appears courtesy of Genex Corporation.*

To our parents
Edward and Ellen Sylvester
and
Aaron and Grace Klotz

Contents

Preface and Acknowledgments

This book came about because of a transcontinental phone call, and like most brainstorms the idea arrived pure, simple, and in a flash: Lynn Klotz, the scientist, and I, the journalist, would write a book whose aim would be to bring into sharp focus the genetic engineering revolution going on around us, a revolution whose impacts we believed would increase exponentially in years to come. We would teach the science related to genetic engineering, for the first time at a layman's level, much as writers and scientists worked in the 1950s to bring the stories of space travel, astronomy, and nuclear physics to the wide-ranging audiences whose lives were to be affected by them. And we would attempt to discuss honestly the risks and flaws of a developing industry, in time for people to come to grips with them.

What followed is typical of a brainstorm's aftermath: hundreds of hours of dogged work, travel, and interviews in person and by telephone with those who are shaping this revolution, exchanges of letters, research in documents, writing, rewriting, perfecting, arguing and rearguing, all required to make a book out of an idea. But I believe and hope that, after all that, we have remained faithful to the idea: that we would detail the magnificent promise surrounding us while keeping the many issues involved in realistic perspective, that we would not let ourselves be carried away and therefore would not lead readers astray, that we would make the book lively and clear at the expense of anything but accuracy.

We felt from the start that if we could satisfy each other's demands in this book, we could produce something of value to all those with an interest in the developments in genetics within the last quarter-century. We believe that should include everyone, whether the interest lies in business, in science, or just in seeing what shape the future is taking. We believe the book will interest those with a great deal of knowledge of the subject, yet be readily accessible to virtually any reader.

The distance from East to Southwest in that original phone call mirrored at least as great a divergence in our own paths. Lynn Klotz and I were roommates at Princeton, and there, in the mid-sixties, we burned many a midnight lamp talking about the exciting developments he was witnessing as the properties of DNA, the life-control molecule, were being discovered. Our wildest speculations were about the kinds of discoveries this book is about, but neither of us foresaw them in our lifetime. Even much of the science of genetic engineering that we discuss at a layman's level here was not known to leading molecular biologists at that time. More startling to Lynn: despite his interest in the subject, he never suspected he would himself be part of a scientific revolution.

Back to that phone call. In the years since college Lynn has been a physical biochemist, for most of those years at Harvard; many of his students and colleagues were and are front-runners in the race to develop genetic engineering. He had called to say he was forming a genetic engineering company, and he touched once again on a subject that already had interested me for some time. During that same time, I have been a journalist—not a science writer, but a writer of all kinds of stories that compelled my interest. The fascination here was not just in the wizardry of this arcane, pure science, though there is plenty compelling about that, but a feeling we shared that in unfolding this story we would be looking at the future as truly as we would by studying the growing patterns of any embryonic living thing.

Putting together the book with Lynn's vital advice and comments, I talked with dozens of scientists, business executives, and government leaders, generally in person, sometimes in lengthy telephone interviews, all taped to assure accuracy in the writing. At times Lynn's own opinions and experiences were important to the story, and in those cases I quoted him directly, as source rather than coauthor. Later, in writing the sections on the science of genetic engineering, Lynn became the principal writer and I the critic and rewriter, and his ability as an explainer of science is, I believe, evident. We designed the science sections to deliver the essentials of molecular biology *as it relates to genetic engineering*; to keep readers from being lost in a whole new universe of particles, functions, and terminology, we have trimmed the science to the very barest skeleton, yet have tried to leave out nothing essential to an understanding of what scientists mean when they speak of splicing genes, achieving protein expression or, that favorite new word, *cloning* things. We coined the term *genefacture* to encompass the new industrial processes harnessing the power of genetically altered microbes, but other than that the terms we use are those already in use by genetic engineers.

This book is about what scientists call "the, cutting edge," the place where science does not stand still for one moment, let alone for the time it takes to produce a book. Yet in one respect, the speed of discovery accounts for an amazingly natural finish to this work—the final section that we call "The Cutting Edge: The War Against Cancer." During the time in which we finished our writing and the book went into production, scientists have been making remarkable discoveries about the nature of cancer, using the same techniques we discuss; and by the time the reader reaches this section, it should be possible for anyone to understand a good deal about what those discoveries concerning cancer viruses and cancer genes mean in the development of disease.

Since most of *The Gene Age* is based on interviews attributed directly in the text, any material taken from published information is cited in notes listed by chapter at the end of the book. These notes also may serve as a guide to further reading, but they do not contain supplementary information—that is, the reader does not need to flip to the notes for explanations in order to follow the discussion.

We are grateful to many people who helped create this book and hope that its quality matches the quality of their assistance. Most are quoted within, but we owe special thanks to John Hunt for his careful reading of the manuscript in its early stages, helpful criticisms, and much-appreciated encouragement; to Dr. Lewis

Cantley for his painstaking and lucid explanation of the latest discoveries concerning oncogenes; to Dr. William Gartland of the National Institutes of Health for providing valuable research documents; and to Jenny Lodge and Suzanne Lawson for the drawings.

Ed Sylvester
Tempe, Arizona
March 1983

THE GENE AGE

chapter 1

Industry Comes to Life

Man invents tools, and the tools change man—so goes the adage. But the change is not instantaneous, and those who watch new tools work magic on their inventors see three phases in the process. First, people use a new technology to accomplish something they've been doing all along, only better. Then they move on to put the discovery to new uses, to accomplish new goals. Finally—the most interesting and hardest to forecast stage—the new technology alters the very shape of the society in which its inventors live, work, and invent. In other words, the discovery produces a change in context, a profound change in human society and social relationships.

In just this way during the waning years of the nineteenth century, the internal combustion engine arrived on the scene as merely a replacement for the horse pulling the carriage and the plow, then became the key to the airplane and submarine. Finally it brought us to a society of suburban living and shopping and freeway commuting, a pattern of life completely unknown a century ago. The engineer-mathematician-venture capitalist who pointed out those phases noted as a measure of their magnitude that his own field, electronic data processing, is really only at the end of phase two. For hints of what phase three will bring, look to satellite communications and space travel, both children of the computer age.

Genetic engineering promises a revolution more far-reaching than that wrought by the computer, one that may bring to a close the industrial revolution that has been the major shaper of our society for three hundred years. Few scientific breakthroughs have been the subject of more overstatement in terms of what can *now* be accomplished. Yet if we attempt to look into the future, around the corner of a new century less than twenty years away, it is impossible to exaggerate the potential of genetic engineering for good and, if misused, for evil.

The creators of the industrial revolution harnessed nature's brute forces to produce steel and other alloys, plastics, and the chemicals that are the hardware of all modern economies. But the industrial hearths demanded coal stripped from the earth, electrical power from dammed rivers, more and more oil. And these hearth fires yielded an abundance of maladies along with the benefits: noxious wastes, an environment fouled by smoke, and the constant threat of shortages.

Genetic engineering, the hearth of the new industrial revolution, promises to deliver the current products of many industries and produce a spectacular array of new ones without this industrial hangover. Energy-intensive furnaces and pressure cookers will give way to sterilized vats in which chemicals will be "grown" as by-products of bacterial colonies, or formed through the intercession of the enzymes that regulate life. And this "genefacture" will be carried out at ordinary air pressure and at the temperature of a fine spring day. Instead of causing pollution, microbes may some day devour landfills and give off valuable methane gas as a bonus, and they may enable us to recover sulfur from pollutants. The forerunners—perhaps even progenitors—of such microbes already exist. An oil-eating "bug" got one of biotechnology's major patents, though its inventors say it does not operate on a scale large enough to have practical uses.

Scientists have already been able to turn microbes into millions of tiny factories producing drugs, alcohol, and chemicals, and astonishing forecasts are being made for genetic engineering's future contributions to health, pharmaceuticals, and agriculture. But there still remains what one top executive calls "the agonizingly difficult transfer of technology" from laboratory to marketplace, and the speed with which genetic engineering sweeps through our lives and our society will largely depend on how long this transfer takes.

Genetic engineering is a fusion of pure science and economics, of laboratory and market. Never before has such a hybrid offered so much in so short a time, or attracted such a clamorous response from the investment community, the public, and even scientists themselves.

Most of the attention has focused on powerful new drugs, vaccines, and diagnostic techniques, both because they are already beginning to reach the public and because any discovery that affects our health is important, personal and, for those reasons, dramatic. But in the long run many in science and business believe that genetic engineering in industrial chemicals and agricultural plants will change the shape of our lives more profoundly than its medical applications will. In fact, the only application that holds greater promise *and* threat would involve engineering permanent changes into the genes of living people. A year ago science forecasters seemed unanimous that such changes were in the distant future. Then, as often happens, scientists shattered the predictions. One group got growth-inducing genes from rats to *function* in mice —the first time genes from one species have been made to operate in another.[1] Then, a second group corrected a deficient human gene that causes a severe blood and bone disorder called β-thalassemia. Neither of these achievements will allow us to correct deficiencies in living people, but they bring that day radically closer.

Not surprisingly, this field, while bursting with promise, is a battlefield: among executives in hot pursuit of profitable products; scientists in universities concerned with the effects such practical pursuits will have on academia; members of Congress concerned with the uses of public money; and clergy who fear we have again advanced intellectually without sufficiently considering the ethical implications. These are important debates, to be discussed as this story gradually unfolds—a story about what scientists call "the cutting edge"—the very fine, sharp edge of knowledge and scientific advancement—and the "real world" in which we live and work.

For all that has been written, many people are understandably confused about what genetic engineering is and how it is accomplished. By the end of this book we believe we can cut through the thicket and answer most of these questions. To begin, the key to genetic engineering as well as the "crown jewel" of molecular biology and the secret of all heredity is a long-chain chemical molecule called deoxyribonucleic acid—DNA. This molecule, found within every living cell, controls the development and function of all life on earth. Whether the cell is one of billions making up a complex human or the single cell of a bacterium, it contains at least one molecule of DNA. DNA functions much like an architectural blueprint, instructing the cell to produce the proteins essential to its survival. When scientists or doctors speak of "genes," they are referring to something which as a physical entity is *nothing more than a region along the DNA molecule.*

Less than ten years ago at Stanford University, Dr. Paul Berg discovered a way to splice genes (that is, DNA regions or fragments) from one kind of virus into the DNA or another kind of virus—he "recombined" the genetic information by recombining DNA fragments. That was the first *recombinant DNA* experiment. His plans to go one step further and insert viral genes into the DNA of the bacterium *Escherichia coli* (*E. coli* for short) began the long controversy over potential risks to humans in such experiments, as we'll see.

The controversy over recombinant DNA began within the biological science community but spread much farther, and although many of the scientists' initial questions have now been answered, others remain. The major fear was based on the fact that *E. coli* is one of many types of bacteria that dwell harmlessly in the human intestines: What if an experimental strain were turned into a deadly pathogen and escaped from the lab? Would we face a plague unmatched in history? In this debate Berg himself favored extreme caution. Guidelines were set up by the federal government, and eventually some of the initial caution was seen as overreaction.

The way had been opened for genetic engineering. Scientists immediately realized that if you could instruct a rapidly reproducing *E. coli* to make a virus protein along with the proteins it normally manufactures for its own livelihood, then you ought to be able to instruct it to make a valuable protein such as human insulin. Or human growth hormone. Or interferon. Engineering foreign genes into *E. coli*, yeasts, and other fungi so that they would instruct these hosts to make proteins or other chemicals of great human value has been the consuming effort in the years since. Making the products we discuss in this chapter involves the basic process of inserting the foreign DNA carrying instructions for a valuable enzyme, hormone, or other protein into the DNA of some other organism, so that the host organism makes the desired protein at the same time it makes its own. That's our definition of genetic engineering.

Once you've got the redesigned bacterium, there's nothing left to do but grow a whole colony from that single parent in the process called *cloning*, a procedure that exploits the principle that bacteria reproduce by division of a single "parent" cell into identical "daughter" cells.

The keys to high performance in genetic engineering can be stated simply: rebuild a bacterium's DNA so it makes the right protein in the largest possible quantities, and grow a large colony from that single parent. As rapidly as that can be done on an industrial rather than laboratory scale, the promise of genetic engineering will be realized.

But what a realization that should be! The U.S. Office of Technology Assessment (OTA) predicted in mid-1981[2] that genetic engineering could displace standard manufacture of products valued at $27 billion over the following twenty years. That's the approximate size of the American organic chemical industry now. Furthermore, the man who provided much of the supporting data for that forecast now believes he erred on the conservative side. Dr. Leslie Glick, president of Genex Corporation, based in the Washington area and one of the earliest gene-splicing companies, was a key consultant for the OTA study. Based on "good, hard data," Glick told us he believes the market for genetically engineered products will be close to $40 billion by the turn of the century—less than two decades from now. His breakdown: $14 billion worth of existing products (half of them now produced from petrochemicals) will be made through genetic engineering, and another $26 billion in brand-new products will be on the market.

Of course, such forecasts are full of variables, and there is by no means universal agreement about whose figures will prove true. Dr. Orrie Friedman of Collaborative Genetics in Waltham, Massachusetts, another major gene-splicer in this field of small companies, believes certain projected markets are extremely inflated— often by accident. For instance, interferon only two years ago was valued at $10 billion a *pound* in a Harvard medical research newsletter, but that was because it was an ultrarare chemical that could only be culled from human cells.

Friedman predicted then that when interferon became available in quantity through genetic engineering the price would drop, even though the protein is still unproved as a drug. Friedman's words were prophetic. By the middle of 1982, interferon was available in quantity *free* to any researcher—its many manufacturers hoping an investigator would unlock a secret that would make manufacture profitable. But, Friedman continued, even if interferon does prove to be a "tremendous wonder drug," the world supply for a year "would be on the order of pounds, and the price will drop drastically because so many people will make it."

Similarly, scientists at Genentech have installed a 750-liter fermentation tank—tiny by the standards of many existing industrial fermenters but big enough, the company noted, to culture enough bacteria to make the entire world supply of human growth hormone, a proven drug in the treatment of dwarfism and one believed useful in speeding wound-healing. Genentech, in South San Francisco, is one of the earliest gene-splicing firms and probably makes the largest number of products of the new companies.

That isn't to say such drugs will be unprofitable. Pharmaceuticals, especially vaccines, are almost always given in tiny doses or shots, and the unit costs are correspondingly high. The difficulty in figuring the relationship between availability and price, however, is an inevitable weakness in any forecast.

No matter. If some of these forecasts prove high, others low, their magnitudes still can help us map out the scope of the genetic revolution on its three major fronts: medicine, industry, and agriculture.

MEDICINE

Vaccines—You sneeze. The flu is coming on. That means viruses have gotten into your body and attached themselves to a few cells and worked their way inside. Viruses may be the world's simplest "organisms," if organisms they are: although they come in a multitude of forms, they are nothing more than DNA or RNA (ribonucleic acid, a genetic molecule very similar to DNA) wrapped in a protein coat. But watch them in action. By some type of chemical trigger, the viruses insert their DNA into your cells. Now, in nature's own version of genetic engineering, the viral DNA takes over the reproductive mechanism of your own cells, forcing it to reproduce viral DNA along with your own. Hosts of new viruses are now made in your cells, usually budding out from the cell surfaces, then spreading to carry the invasion. You don't feel a bit well. But you're not whipped yet.

Viruses have certain surface markings as easily readable as an invading jet's are to an air observer. Antibody cells constantly swimming in your blood recognize these invaders, or antigens, and attach to them. Antibodies are very specific: each type recognizes only one type of invader. This initial "interception" of the invaders also appropriately triggers a general mobilization—the production of vastly more antibody cells, which secrete huge amounts of antibodies into the bloodstream. They also attach to invaders and literally sink them out of the blood. The war is on. Eventually the antibodies win: you get better.

This or a similar scenario unfolds at the onset of all viral infections, and the antibody interception marks the starting point for cure and prevention. In order to vaccinate against a particular flu, for example, doctors take a batch of the virus causing that flu, kill it, and inject it into the bloodstream. If all goes well, the antibodies recognize the surfaces of the *dead* virus and sound the general alarm, producing more antibodies. Thus the antibodies are massed in advance to repel future invasions of the same flu virus before it gets there. Polio vaccine works in much the same way, and it protects for life.

But there is an inherent hazard in all so-called killed-virus vaccines. If the virus by accident is not completely deactivated, patients may get the disease instead of becoming immunized against it, an occurrence frequent enough to make many such vaccines unpopular for general use around the world. Further, there are some dead viruses to which the body's immune system for some reason will not respond, so antibodies don't multiply in preparation for attack. Killed-virus vaccine serves no purpose against those diseases.

Genetically engineered vaccines could solve these problems, because they are not composed of killed viruses but only of virus *parts*. Ultimately, perhaps, the parts may consist of only those surface markings that trigger antibody response. Such vaccines would never contain the whole virus, so they would be completely safe.

The first genefactured vaccine on the market may be that for hoof-and-mouth disease. It is already in production by Genentech, which created it with support from the U.S. Department of Agriculture, and should be ready for overseas sales in the mid-eighties. There is little potential market in this country because hoof-and-mouth disease has been under control here for many years, but it could save hundreds of millions of dollars a year in Third World countries—those which can least afford the cattle losses.

To get some idea of the extent of the disease, consider that over one billion doses of conventional vaccine are administered every year, often to no avail. The vaccine costs 20 cents a shot, but genetic engineers believe this can be cut by 50 to 100 *times* using recombinant DNA manufacture of one of four of those "signal" proteins on the virus's surface.

A Genentech spokesman said recently that progress in marketing has been slower than forecast, because a whole cluster of *slightly* different viruses causes hoof-and-mouth disease, and the company has had to reproduce a vaccine to combat all of them, a painstaking job.

Other viruses frequently change those surface markings, so new strains go unrecognized by the antibodies, forcing doctors to constantly chase down new vaccines against the everchanging "bugs." Tracking human flu viruses as well as animal microbes should become vastly easier via genetic engineering.

Potentially far more important, a vaccine against hepatitis has reportedly been produced for the first time in history, in marketable quantities and at a modest cost, but it must pass lengthy U.S. Food and Drug Administration tests before it can be sold. One firm recently reported creating a vaccine against malaria, a disease that ravages millions of people in tropical climates, but that vaccine may not even be ready for testing for some time.

*Hormones and enzymes—both simply forms of protein—*are perhaps the largest class of genetically engineered pharmaceuticals soon to be available, the best known being insulin. This hormone that regulates the metabolism of sugar is now being made commercially by the Eli Lilly company under contract with Genentech. Biogen, another gene-splicing company, has developed its own genetic engineering process to produce insulin for Novo, a Danish firm. Lilly and Novo now make virtually all the world's insulin, giving Genentech and Biogen not only a scientific head start but a ready, giant market for their products. Originally expected to be through testing stages and in the hands of diabetics this year in the United States, this new insulin passed stringent FDA requirements in near-record time and has been on the market since late 1982.

The clamor over genefactured insulin has confused some people, who think a cost reduction will result—generally it won't, but the new product will be very pure and most significantly, it will be human insulin. Until now, diabetics have always taken insulin gathered from the pancreases of slaughtered cows and pigs. Both animals' insulin is remarkably close to human, but not identical. Researchers suspect that allergic reactions among some diabetics may be caused by these small differences. If they are right, the availability of true human insulin might solve a major and quite hazardous allergy problem, but initial reports have not answered these questions. Although the ultimate value of the new insulin still is unknown, a Genentech spokesman noted that the early approval had made human insulin the first product of genetic engineering to reach the public.

Wall Street stock analyst Scott King of F. Eberstadt & Co. recently estimated that Lilly's human insulin would bring $11.5 million in revenues in 1983, with about $1.2 million in royalties going to Genentech. By 1985, King forecasts, Lilly's human insulin revenues would represent 75 percent of the U.S. market and be worth $186.8 million, with Genentech's share up to $18.7 million. But here is an example of how unknowns can play havoc with seemingly solid forecasts. King predicts that by the end of this decade a cure for diabetes will be found, potentially eliminating the entire market for insulin.

Firms have already produced testable quantities of human growth hormone (HGH), which stimulates cells to grow and divide. The lack of HGH causes dwarfism and, as noted, because of its role in cell growth scientists believe it may be a major aid in healing wounds and burns. Researchers are also investigating ways to genetically engineer such proteins as factor VIII, lack of which causes hemophilia, and another whole group of hormones called endorphins and enkephalins that have stirred excitement among medical researchers. The latter two are produced in the brain and are thought to be natural pain-killers. In fact, these or similar hormones are thought to be responsible for "runner's high," the feeling of well-being experienced by long-distance runners and joggers. If so, scientists might eventually produce powerful yet nonaddictive pain-killers and anesthetics by recombinant DNA means.

Interferon—No little-understood protein has caused the excitement in research or attracted the publicity of interferon. First isolated about twenty-five years ago, interferon appears to be a natural antiviral agent *inside* the cell, as antibodies are in the blood fluids outside the cell. There are actually nineteen or so similar proteins known as interferons. Scientists have long hoped that one or more would prove to be the long-awaited "magic bullet" for all viral disease, killing viruses with the efficiency that penicillin and other antibiotics wipe out bacterial infection. Early hopes that even cancer would be cured with interferon have not flowered, but interferons are known to work against many monkey viruses, and best guesses now are that ultimately they will counter many viruses in humans—perhaps even the common cold, as current evidence suggests.

The New York Times on May 5, 1983, reported that preliminary studies at the Memorial Sloan-Kettering Cancer Center in New York indicate that genetically engineered interferon may be useful in fighting a rare skin cancer, Kaposi's sarcoma, which afflicts victims of a dangerous immune system deficiency known as acquired immune deficiency syndrome, or AIDS. Interferon was also reported effective in fighting the collapse of the immune system that is fatal to half the victims of AIDS.

How interferons work is still a mystery, but somehow they prevent some viruses' genetic material from taking steps that are vital to their takeover of the invaded cell. That might imply that a simple interferon injection would cure viral infections. Maybe, but interferon seems to work by interfering temporarily with the cell's *own* DNA function—just long enough to prevent the viral takeover. If that's the case, interferon injections might permanently damage necessary cell mechanisms. The answers to these questions could be at hand or years away, for initial reports on its effectiveness have been contradictory. In 1982, four French cancer patients died during large-scale trials of interferon, causing the government to suspend further testing. But the interferon used was conventionally gathered from cells, not made by rDNA. The conventional method yields a very impure product, as low as 0.1 percent interferon; the rDNA product can be 10 percent interferon. Researchers are not sure the deaths can be attributed to the impurity—or whether they can be attributable to the interferon at all.

Researchers in England also reported side effects in patients treated with purified interferon for breast cancer. Many patients registered irregular brain-wave patterns and suffered lethargy and "confusion," but the FDA said these side effects were predictable.[3]

Even if interferon works, the profitability to the dozens of firms making it may be mixed. Recalling Friedman's caution, if the world market turns out to be measured in pounds per year, a few companies may realize a lot of revenue from interferon, but it certainly won't keep afloat all those trying to make it.

Bulk pharmaceuticals—On the other hand, the U.S. Office of Technology Assessment foresees major markets for genetically engineered, *bulk-produced* pharmaceuticals, including vitamins and antibiotics, within ten years. Pound for pound, these drugs don't earn as much for their makers, and they require huge scale-up in capacity of fermentation vats and other equipment, but the markets for them are correspondingly huge. Vitamins alone represent a $500-million-a-year industry, and the OTA forecasts that genetically engineered organisms could soon produce them economically enough to displace current methods. Similarly, sources in the field predict antibiotic production could be aided by genetic engineering in two to four years, marking a major cost reduction in an industry whose annual sales are $4 billion.

But there are cautions here, too. Vitamins and antibiotics are *now* produced via enzymes, without recombinant DNA technology, but in the same kinds of "factory cells" we've been talking about. A long, complicated series of production steps is followed within the cell. Recombinant DNA methods *could* speed up some slow steps, perhaps even radically, but tinkering with such complicated processes could result in no net benefit in some cases because of unexpected side effects. In other words, the potential for gene-facture in this area is high, but the near-term uncertainty is also high.

The OTA provided a clearer forecast for the future of acetaminophen, better known by Johnson & Johnson's trade name, Tylenol. The federal study estimated current manufacturers' costs at $1.325 a pound, but projected that genetic engineering methods could cut that cost to $1.05 a pound. Given ordinary drug price markups, that would mean a decrease in price to the consumer of $5 to $10 a pound—$100 million to $200 million a year for one drug that would, meanwhile, become part of the genetic engineering product "stable." That estimate was made for the OTA by Glick, who developed a plan for manufacturing 10 million pounds of acetaminophen per year from one factory containing two 50,000-gallon fermenters.

Monoclonal antibodies—Interestingly, the first product of biotechnology to affect human health and the fiscal health of the industry in a major way may not even involve recombinant DNA, but a different product of biological wizardry called the *monoclonal antibody*. We described how antibodies are the immune system's army of defenses against many forms of invaders, and how each antibody homes in on one and only one recognized site on the attacker, making it necessary for *multitudes* of different antibodies to be sent out against any given invader.

For some time scientists have been trying to isolate and raise colonies of *single* antibodies. In 1975, two scientists did it. They fused myeloma cancer cells (which multiply extraordinarily fast) with antibody cells. Then they spread the resulting colony so thin that all the cells were widely separated—widely enough that any single cell could be grown into a whole colony separate from the others. This is the process of cloning, to be described in detail later. In this way, the scientists got whole batches of the same (monoclonal) antibody.

A serum of such single antibodies would flow quickly to one and only one site in the body, so scientists hope that monoclonal antibodies can be used to carry drugs, antibiotics, or even cell-killing radiation to specific sites of infection or tumor growth, leaving healthy tissue unaffected. That could eliminate the side effects of much existing radiation- and chemotherapy.

Diagnostics represents a major area in which monoclonal antibodies should quickly replace existing techniques. Because they recognize specific antigens, these antibodies could be used in test kits for pregnancy, cancer, and virtually all infectious diseases. The kits would be inexpensive, safe, and far more accurate than those now available.

Fermentation—Because the discoveries of genetic engineering are so recent, the aura of newness tends to extend to everything involved in biotechnology, but much of enzyme manufacture and other fermentation is ancient technology. Alcohol was produced by the Sumerians and Babylonians before 6000 B.C. Egyptians were using brewer's yeast to leaven bread in 4000 B.C. Production of purer alcohol through distillation of fermented grain was common in most parts of the world by the Middle Ages.

Louis Pasteur, identified by most of us with his work on rabies vaccine and "pasteurization" of milk, did much of his research to aid the French wine and beer industries. His studies of yeast revealed that this organism ordinarily respirates, like all other living things, using oxygen to burn food and giving off carbon dioxide as waste. But unlike most living things, Pasteur found, yeast can live without air by shifting metabolic gears into an entirely different way of deriving energy from food—anaerobic (without oxygen) fermentation. Though not nearly as efficient an energy producer as respiration, this airless fermentation gives yeasts an edge in surviving, because it is carried out in an airless environment that would kill most organisms. The major waste product of fermentation: ethyl alcohol.

It might seem that alcohol production is the only ancient use of fermentation technology and therefore an anomalous example, but not so. Writing in a special *Scientific American*[4] issue on industrial microbiology in September 1981, Douglas Eveleigh noted that bulk fermentation is "not just a new field of entrepreneurial activity; it is a well-established factor in the world economy, responsible for a current annual production valued at tens of billions of dollars." Antibiotics certainly constitute a major product. The four most important groups of antibiotics—penicillins, cephalosporiums, tetracyclines, and erythromycin—had bulk sales in 1978 of $4.2 billion. They are all produced by bulk fermentation.

These drugs are natural metabolic products of the microbes from which they sometimes derive their names (penicillin, for example, is derived from *penicillium* mold). Vitamins are known as primary metabolites because organisms create them as an essential part of their life-support processes; they are equally vital to humans, but we cannot make them and must get them from food. Antibiotics and other chemicals such as alcohol are known as secondary metabolites. Microorganisms make them either as waste products or—in the case of the "wonder drugs"—possibly to attack other bacteria that attempt to invade their living grounds.

Forcing such bacteria to mutate so they "overproduce" their natural antibiotic in greater and greater yields has been a practice in the pharmaceutical business for decades. For example, after Alexander Fleming discovered the antibiotic properties of penicillin in 1928, it took a dozen years for scientists to put the discovery to work and to force production levels needed for practical use of the drug. In fact, it was an American scientist working for the Department of Agriculture who developed the bulk fermentation process used in the first commercial production. Now, companies can get 10,000 times that first production level.

But industrial microbiology already encompasses more than drugs. The amino acid industry is worth $1.7 billion a year, even before genefacture streamlines its methods and improves products. Amino acids, the building blocks of proteins, are valuable nutritional supplements for animal feed, but the best known of them, glutamic acid, is eaten by humans. Better known by the chemical name of its salt, monosodium glutamate, MSG is consumed as a meat tenderizer and flavor enhancer at the staggering rate of 300,000 tons per year, and it too is produced via fermentation.

The relationship of industrial fermentation to recombinant DNA is as close as hand in glove: the bacteria or other microorganisms redesigned through genetic engineering must then be grown in huge quantities in order to mass-produce the proteins they are making, and generally that is done by a form of fermentation.

INDUSTRIAL CHEMICALS

For years the organic chemists working in industry to develop plastics and other products have been aware of an embarrassing inefficiency in their pathways from starting ingredients, or "feedstocks," to the cooked-up end-products. In the *Scientific American* issue mentioned earlier, microbiologist Eveleigh quoted from the long-standing "organic chemist's ode":

> Lord I fall upon my knees
> And pray that all my syntheses
> May no longer be inferior
> To those conducted in bacteria

The prayers are being answered—no small irony here—with the newfound cooperation of the bacteria themselves. The greatest potential for biotechnology in the long term may prove to be production of major industrial chemicals that are now the product of brute-force technology. Essentially, that means replacing energy-hungry inorganic catalysts with enzyme catalysts of incredible specificity. Eberstadt analyst King pointed to a huge chart on his office wall displaying the known enzyme chemical pathways. "That's what biotechnology is all about," he said. "Whatever major [activity] is going to come out will involve a better understanding of enzymes."

Enzymes act principally as biological catalysts of chemical reactions. A *catalyst* is a substance that speeds up specific chemical reactions, but remains unchanged itself at the end of the reaction; for that reason, the same catalyst can be used again and again. Effects of catalysts can be dramatic. Some chemical reactions that could take longer than a lifetime to occur without a catalyst are completed in a few seconds in the presence of the right one. Catalysts are a mainstay of chemical manufacture, but most catalysts now used are not enzymes. With the availability of cheap enzymes through genetic engineering, however, much of the chemical industry may "go enzymatic" by the end of this century.

To oversimplify the valuable specificity of enzymes, imagine that every time you mix chemical A with chemical B, you get a giant batch of C, which may or may not be useful. But just by chance, a few undetectable molecules of another *very* useful chemical, D, are formed on the rare occasion that A and B join up in a slightly different form. Small quantities of other AB combinations will form by chance too. Chemists work to find ways by which, sometimes at high temperature and high pressure with the help of a catalyst to trigger the reaction, they can consistently make batches of D; because once formed, the molecule may be very stable. *Enzymes* are biological catalysts, and they can carry out the highly *selective* manufacture of given D molecules. If an enzyme makes D, it makes D alone, without by-product. That's what we mean by *specificity*. The synthetic chemical industry, from plastics, adhesives, dyes, and solvents to food additives, is based on the properties of catalysis: muscling molecules into different shapes than they take in nature, or producing them in much larger quantities than they are naturally found.

The association of industry with power and brawn is not misplaced. Molecular bonds of industrial feedstocks are often strong. Breaking them requires a lot of energy, which is lost as heat when the bonds are re-formed into their useful combinations. What is left behind often includes polluting residues that are difficult or impossible to get rid of.

By contrast, the enormous variety of molecules that compose the human body are basically built out of carbon, hydrogen, oxygen, and nitrogen. The combination of these elements into various molecules is determined almost entirely by the catalytic action of enzymes. That's an indication of how important this class of proteins called enzymes is, and how powerful. Many of the molecular bonds of living things are also very strong.

Enzymes mediate the breakdown and reshaping of all these chemicals in ways which require much less energy. They are nature's answer to the blast furnace and the ten-ton press. That's why molecular biologists believe that the opportunity to put enzymes to their own uses will change the face of industry, as they replace energy-hungry inorganic catalytic processes with energy-efficient enzyme catalysis of incredible specificity.

Such a change will not represent the first exploration of producing chemicals via biotechnology. Pasteur's work led a German biologist named Eduard Buchner to discover cell-free metabolism. That is, he learned how to obtain the products of cellular metabolism without anything but a mix of enzymes and their "food" as feedstocks. During World War I, Germany used industrial fermentation to produce glycerol for explosives after the British blockade cut off supplies of the usual feedstock. As much as a million pounds of glycerol a month were produced by fermentation. Meanwhile, the British were short of acetone, another vital ingredient in munitions manufacture. Chemist Chaim Weizmann developed a method of producing acetone biochemically (he later became first president of Israel). Weizmann's method for producing acetone was used for many years after the war, losing out finally to processes based on petroleum—then so cheap and in seemingly inexhaustible supply.

Some genetic engineering firms believe enzymes derived from microorganisms will soon become tools in the $20-billion-a-year plastics industry. Several plastics are made by using a class of base chemical known as alkene oxides, now derived from petroleum.

A scientist at the Cetus Corporation in Berkeley has devised an enzyme method for making these oxides, and the company believes that enzyme production of just one of these—propylene oxide—will begin before the decade's end and eventually will be worth $2 billion to $3 billion[5] a year. Propylene oxide is just one of the many chemicals whose homely names belie their wide-ranging value as feedstocks—this one for urethane foams and ingredients for other plastics, solvents, fumigants, batteries, and even explosives.

Alcohol—Of all the chemicals that can be made by genetically engineered organisms or via the enzymes those organisms produce, perhaps none is more immediately valuable than plain ethyl alcohol, that ancient beverage and century-old antiseptic that we have recently come to consider seriously as a fuel. We tend to ignore the fact that alcohol is already the most important industrial solvent after water and a major starting material for making industrial products. Synthetic ethyl alcohol has a market value of some $315 million a year. Naturally fermented alcohol—which is used for beverages as well as for industry—has a market of more than $600 million a year.

Ethyl alcohol, also called ethanol, can be made from waste paper, wood pulp, and other agricultural scrap. Such a conversion also has a social benefit—removing waste—that can defray costs of what might otherwise be an uneconomical process. The current production *cost* of alcohol from scrap, according to a U.S. Army study from Natick, Massachusetts, is $1.38 per gallon.[6] In February 1983 gasoline *prices* fell to 99.9¢ per gallon, and were projected to drop even further, so alcohol is not yet competitive as a fuel. But consider the production process. A mash of cellulose-containing products is treated with a variety of enzymes called cellulases, to yield glucose. Yeast is then used to convert the glucose to alcohol. The army study sets the price of the cellulases, which are made by a fungus, at 57 cents of the $1.38 cost per gallon. But some scientists believe that genetically engineered bacteria could manufacture *better* cellulases *faster* than the fungus does, so that enzyme costs might be cut from 57 cents to 10 cents per gallon of alcohol. That could cause a radical change in the use of alcohol and the cost of the products in which it is a component.

Further enlivening the possibilities for commercial alcohol gene-facture, biologists are trying to accomplish fermentation with a class of bacteria called thermophiles, which thrive at temperatures of around 158 degrees Fahrenheit (70 degrees Celsius). That temperature is not far below alcohol's boiling point. Much of the cost of alcohol production is in distilling it, and that means heating up the mash so the alcohol boils off. Currently, the alcohol-producing yeast die off at the high temperatures needed for distillation, so they have to be regrown for every batch of alcohol produced. Alternative: genefacture the alcohol in heat-loving (thermophilic) microbes at high temperature, so that with only slight suction the alcohol would boil off as it was being *fermented*. The thermophile would remain alive to produce more alcohol indefinitely.

Much ethyl alcohol is now derived from corn or grain, and both ethical and practical problems crop up in discussions of fuel alcohol. It has been estimated, for example, that to run only 20 percent of America's cars on alcohol would require turning the country's entire corn crop in a *record* year into alcohol. Analyst King scoffs: "There will never be enough alcohol to use as a serious fuel unless we give up eating." The ethical qualm of turning a hungry world's food supply into fuel can be resolved by using agricultural waste, but the staggering scale of fuel-alcohol projects remains a problem. And there are other, technical problems as well.

For example, in the July 1981 issue of *Science* magazine, four scientists analyzed the use of "biomass"—any mass of organisms or their debris—as a source of chemical feedstocks.[7] They pointed out the advantages of reduced dependence on imported oil and sharply cut atmospheric pollution, and noted that existing industries would be willing buyers; so gene-splicers would not be at a loss to sell their product. But the scientists also warned that the transfer of the new technology does not look at all simple. The current method of producing ethylene, for example, yields propylene, another valuable chemical, as a by-product. If we want to make ethylene exclusively from biomass, a way will have to be found to make propylene economically as well. They estimate that alcohol will not enter the feedstock market until its cost has been cut fivefold. Impossible? No, but uncertain. However, the social "redemptions" of using biomass may make some economic loss worthwhile, especially as part of a waste-conversion project.

Still, we must not overlook the fact that alcohol is already used as both a fuel supplement (in gasohol) and an industrial solvent. Genetic engineering may begin to make itself strongly felt as one producer even before it replaces other methods. Similarly, most of the industrial enzymes may soon have economical, revenue-producing uses, even if they do not reshape a whole industry for some time.

Glucose isomerase is the key enzyme in converting glucose—a relatively unsweet sugar found in both grapes and blood—to fructose, a very sweet sugar. It would take less of such a super-sweet sugar to bring a beverage or commercial food to the right taste level, so most genetic engineers believe pure fructose would be much in demand. Even now, "high fructose corn syrup" (HFCS) is used extensively in industry because, even though the overall mix of sugars in it only brings HFCS to the sweetness level of table sugar, it is far cheaper. Coca-Cola recently announced that as much as half its sugar consumption for noncola drinks would soon be of HFCS.

High fructose corn syrup is made from cornstarch. It now costs a major supplier of glucose isomerase from $2 million to $3 million a year to make its product. One molecular biologist believes that he can soon genefacture a more efficient and more stable glucose isomerase costing one-tenth the current amount, and that this would translate into $2 million a year in enzyme savings.

Similarly, far cheaper and better genetically engineered alpha-amylase (α-amylase) and glucoamylases, required in the early manufacturing stages of both HFCS and ethanol made from corn, could continue cutting the cost of those products immediately and give an added boost to their markets. Even without that cost reduction, the use of HFCS is expected to double within a couple of years because of a shift from dependence on imported sucrose (table sugar) to corn, of which the United States is the major grower.

Other industrial products—The OTA study predicts that genetic engineering will be used to convert sewage, landfill garbage, and agricultural waste to methane within ten years, although all methane will not necessarily be genefactured. However, the current value of industrial methane, a major feedstock that ultimately could be made by biotechnology, is $12.57 billion. The OTA forecasts that some of the hydrogen and ammonia needed for major industrial production could be produced through genetic engineering within fifteen years. More recent predictions based on new scientific work indicate that this sewage conversion can be accomplished in two to four years.

GENETIC FARMING

Future changes in the way we eat, farm, and think of crops and cattle through biotechnology are staggering in scope, but they are still a few years away. In this area, many if not most of the problems and uncertainties are scientific rather than economic. But the form those changes take should be the one we've discussed since the outset: first, replacement of the way we now do things with new ways; later, new inventions; and finally perhaps even changes in context, one of which we'll suggest at the close of this chapter.

In genetic engineering's first uses, instead of merely cross-breeding crops to grow taller or increase their yields, or crossbreeding cattle to give more milk or yield more meat, we may genetically engineer them to do so. Plant and animal breeding are laborious, time-consuming, and very expensive. We may be near the time when crops can have vital human nutrients that they lack spliced into them. For example, scientists are trying to splice into corn the genes calling for high production of the amino acid l-lysine, which humans need. Since corn is a major staple in some Third World countries, improving its food value would boost nutrition without requiring increased supplies.

Modern farming is harnessed to nitrogen fertilizers, which require huge amounts of petroleum products in their production. Farmers in the United States spend $1 billion a year just to fertilize corn, for example. It all seems like a mistake of evolution—fertilizer is needed only because most plants cannot "fix" nitrogen in their root systems to manufacture key protein-building amino acids, even though nitrogen makes up the bulk of the air around them.

But evolution played a lucky trick on some plants. Legumes like beans, peanuts, clover, and lupine need no chemical nitrogen source and not only grow without fertilizer, but enhance the soil for anything planted after them—a fact that has been known since Roman times. The benefactor is actually a bacterial infection;

rhizobium bacteria infect the roots of legumes in a classic example of symbiosis. The bacteria get their nourishment from the plants, then give the plants nitrogen-rich ammonia in return. Some other plants are also able to fix some nitrogen through the intercession of other bacteria, but none so far as successfully as legumes.

It is often noted that the entire food cycle ultimately depends on plants, which get their energy directly from the sun and use it to build sugar in the photosynthesis process. Animals lack this ability, and must therefore eat energy-giving substances, such as plants, but at the same time they derive many needed amino acids from this food. On the other hand, plants cannot get nitrogen-based amino acids they need through photosynthesis. That's why the *rhizobium* infection is so useful, providing the nitrogen foundation for the amino acids, and why a major goal of "agrigenetic" engineering is to teach plants to fix their own nitrogen.

Biologists also hope to splice in genes that increase the efficiency of plants in converting sunlight to sugars, which could lead to vastly increased plant yields without an increase in fertilizer demand.

Tinkering with a plant's genetic information could produce a dazzling array of agricultural products and benefits, including crops that are able to grow in salty environments. That would mean food crops might be raised on presently nonarable land such as alkaline deserts. Some plants can already live in such deserts, and scientists think that within a few years they may be able to splice the genes leading to that ability into food crops. Further, plants may be engineered to produce natural pesticides, cutting food production costs and eliminating a major health hazard and cause of pollution.

Futurist Alvin Toffler[8] has pointed to the exponential rate of change that confronts us in the twentieth century—a rate of change that makes present methods of prediction obsolete. Applying a similarly rapid rate to genetic engineering, one can imagine some day creating an edible plant that would grow in salt soil without fertilizer, be extremely efficient at photosynthesis, and have other qualities that would make its waste parts ideal for conversion to alcohol. This would be a good example of "synergy," in which advances reinforce one another to accelerate even the rate of change.

Problems?—In detailing some of the difficulties involved, Winston J. Brill, writing in *Scientific American*,[9] divided plant genetic engineering research into three strategies. First, scientists might find more microorganisms that *naturally* help plants grow, produce them in quantity, and feed them into the soil (perhaps with genetically engineered improvements). Or, they can grow individual plant cells in cultures to increase their genetic mutation rate and screen for superior traits. Finally, they can attempt to engineer new genes into plants as they have in bacteria. Brill believes this last will be the most difficult. In fact, he believes it may ultimately prove easier to manufacture some needed plant proteins *in bacteria*.

Some scientists disagree with Brill and believe that the genetic engineering of nutritional additives for crop plants is not that far off. Professors Fred Ausubel of Harvard and William Orme-Johnson of MIT believe that inserting beneficial genes for proteins needed in human and animal nutrition should be possible within a few years. Further, they believe that the even more difficult problem of nitrogen fixation, which involves engineering seventeen genes into plants so they can fix their own nitrogen without *rhizobium*, will be cracked within five to ten years.

But such experiments certainly have built-in difficulties. Brill noted an experiment in which he and colleagues introduced mutant *rhizobium* bacteria into a soybean field to enhance nitrogen fixation. They failed, but the failure was instructive. The lab-designed bacteria could not compete with the natural *rhizobium* strains that had evolved as the fittest to survive, even if evolution had not made them best at delivering nitrogen to plants. "We had encountered precisely the effect that distinguishes agricultural practices: the effect of natural populations . . . in the uncontrolled environment of the open field." As we'll see, these results can also play havoc in industrial fermentations, and the relative weakness of engineered bacteria against natural ones is also a major reason why scientists believe early fears of creating new plagues were generally unfounded. The trick here might be to genetically engineer an already-competitive natural *rhizobium* to be better at fixing nitrogen.

Engineering in animal DNA offers huge technical difficulties, but it raises some serious social questions as well. Are we prepared, as one writer put it, to see a field of six-legged sheep or chickens, created to maximize leg of lamb and fried chicken production? In response to such comments, Dr. Thomas E. Wagner[10] of the

University of Ohio has said that Holstein cows might already qualify as such unnatural animals. They are milk machines, the result of years of painstaking crossbreeding for increased milk production. Wagner's own research tends not toward the creation of freak mutants, but toward the acceleration of the natural breeding process. He recently introduced a gene from a rabbit into the genome, or gene complement, of a living mouse; he sees this as a major step toward creation of "three-parent cattle," in which offspring would get desired traits from an extra parent whose genes would have been placed in the fertilized egg.

TOOLS CHANGE MAN

Before genetic engineering ever produced its first major product, it had an explosive impact on science, business, and the very university system that nurtured it. This book is about those varied impacts and their different forces, as well as about genetic engineering itself. Before a dollar was realized from genefacture, an array of companies had lined up with expectations unmatched since the days of land rushes, and a variety of corporate structures as different in type, size, and plan as could be found in any collection of Oklahoma "Sooners."

Major drug houses and chemical and energy companies have poured hundreds of millions of dollars—by some estimates $1 billion—into an industry that still has slim profits. Like loners on horseback, venture capitalists plow scattered millions into small firms ranging from elite partnerships of science's Nobel laureates all the way down to mere mailing addresses representing only, in the words of one entrepreneur, "a man and a boy." Other firms have tried to bridge the gap between university and business, setting up nonprofit foundations to spur scientific research, while contracting for purely practical work they see leading to profits. Several firms have gone public, with mixed ambitions and mixed results. Each has a different notion of what will bring success in the form of a balm for humanity and enormous profits for itself. Yet this very diversity boosts the chances for a successful future of the industry even as it increases the likelihood that some firms will founder.

Already there have been casualties. Stock prices have risen wildly only to plummet for reasons difficult to assess. A major corporate undertaking, DNA Science, Inc., sponsored by the prestigious E. F. Hutton Co., fell apart before it got off the ground, again for reasons not clear.[11] Something of the firm remains on paper, but its original missions, fueled by a planned investment of some $35 million, have been canceled.

E. Russell Eggers, the president of DNA Science and the former chief executive of both Bendix International and the Loctite Corporation, said in an interview before the bottom fell out, "We're up against an agonizingly difficult transfer of technology" from laboratory to marketplace. "When you mention competition, most of the new companies talk about each other. But they are not competing with each other and won't be for a long time to come. They're competing with the old technology. It's time-tested, its pathways and steps and incremental costs are known down to the penny."

That transfer of technology does not come when science or the new technologists and business people are ready, Eggers says, but when "the factors that underpin the old technology begin to change." Like the *Science* authors (Palsson et al.) cited earlier, former Rhodes Scholar Eggers believes those factors *are* changing, but not as quickly as many would like. Some see the failure of DNA Science to get off the ground as the beginning of a long-predicted "shake-out" in the industry, when the end of easy money would cripple all but a few firms with either heavyweight and firmly committed backers or enough investment money in the bank to survive. More recently, two other firms have declared bankruptcy—Southern Biomedical Laboratories and Armos, Inc.

Part of most firms' *modus operandi* involves tapping the best academic scientists for research, but they must cope with complaints that this commercialism is costing universities their better scientists. DNA Science had planned to create spin-off companies in scientists' university towns, to offer them the chance to remain in academia while still enjoying the fruits of the "bio-boom." Many similar structures have been planned by others but, as we'll see, the controversy goes on. A range of critics from scientists to members of Congress see the boom as a threat to a form of free intellectual inquiry that has been a mainstay of Western culture since the Middle Ages. Who will prove right remains to be seen.

In forecasting the economic future of genetic engineering, at least, Genex's Glick believes the doomsayers have read the omens all wrong. Even if a shakeout comes, he does not believe it will slow genetic engineering in its move into major markets. His forecast that $40 billion in business would flow to the new industry within twenty years was based, he said, on an analysis of $125 billion in products studied to see where biotechnology could cut costs.

Answering those who predict long time lags before cheaper production will be achieved, Glick noted, "The conventional wisdom in 1977 was that the first [human] insulin *molecule* would be created in the lab in 1982. It was done in 1978 [and was marketed in 1982]. At the same time, the conventional wisdom said the first interferon molecules would be cloned in 1987. That occurred in 1979."

He went on to point out that sequencing gene fragments— figuring out what order of bases (the alphabet of DNA, which we'll discuss later) occurred in a given gene—was an undertaking of staggering proportions in the 1960s. It took five scientists as long as ten years to sequence DNA fragments 200 bases long. In the 1970s it took five people one year to sequence fragments hundreds of times that long, and today, one person could figure out a longer fragment in less than a year.

His point: The entire history of modern technology is one of forecasts being outstripped by performance. Electronic calculators in 1970 added, subtracted, and divided, measured one foot by one foot, and his own early model cost about $500. Now for under $200 you can buy one that carries out so many functions that its ancestors would have filled a good-sized room. Glick may be one of the more optimistic of the gene-splicers, but he believes that commercial manufacture of alcohol from waste cellulose will be economical in this century, including the required vast scale-up of production capabilities.

New discoveries accelerate synergistically, each boosting the others' impacts. That, at least, is one theory of how quickly tomorrow will come. But regardless of speed, surely no means of reshaping the future offers more intriguing food for thought than the reshaping of society by its own inventions, and as a hypothetical example of how this might happen in the Gene Age, we offer a case posited by the *Science* authors, Palsson et al.

Genetic engineering may produce industrial feedstocks with great energy savings, but that is only phase one in the progressive change tools bring to their inventors. Of at least equal importance farther down the road is the fact that these new feedstocks might be grown. Agriculture is not only labor-intensive, it is unskilled-labor intensive. The agriculture needed to feed manufacturing plants would not have to be either vast in scale or centralized, but could be dotted over wide areas in many countries, as could the manufacturing plants for the feedstocks. It will be ironic—but a happy irony—if this most sophisticated of all sciences could lead to a system that would bring major rewards to unskilled farmers in underdeveloped countries by allowing them to remain farmers, but to reap some of the rewards of industrial society by growing the raw materials of industry. That kind of change would mark the third phase of a bio-revolution, the change in the shape of society.

Genetic Engineering: The Science I

When scientists say that DNA is the secret of life, they mean that the long, yet awesomely simply built molecule contains the blueprint for each particular living thing in which it is found, a blueprint not only for creating cell parts out of the building blocks we obtain from food, but for assembling parts into the whole and for maintaining the organism in its living state.

The surprisingly simple relation of DNA's form to its function, which we will discuss shortly, has awed scientists since those who discovered its shape first marveled at it thirty years ago. In science as in art, the truly powerful concepts are those which emerge with clean-lined simplicity from dissonance and complexity—and there's plenty of complexity to be smoothed out in learning how molecular biologists have harnessed the power of DNA to turn billions of cells into factories for chemicals and drugs.

In function, DNA is similar to the master blueprint for a planned-community of houses. But it's not a drawing. This blueprint is a set of instructions communicated in a unique code whose "alphabet" uses only four submicroscopic units as letters. Billions of these molecular code-letters could be strung out end to end on a postage stamp.

In this chapter, we look at how the instructions contained in DNA are put into action in the living cell, in particular, in the organism *Escherichia coli*—*E. coli* for short. This simple, one-celled bacterium has long been the favorite subject for microbiological study. We concentrate on those aspects of an *E. coli* cell responsible for reading DNA language and then carrying out the instructions telling the cell how to grow, divide (that is, produce a living copy), and maintain itself in a healthy, living state.

Virtually everything we talk about here applies to the human body and its cells as directly as to bacteria or any other living organism. Our own DNA replicates in much the same way as the DNA of *E. coli,* and with the assistance of enzymes our DNA orders the construction of our own body-building proteins using the very same code-language. Among the many wonders of molecular biology is the fact that the genetic code is universal. No Tower of Babel here: the same "words" that instruct the *E. coli* to add another particular amino acid to a protein chain would order up the same amino acid in a honeybee or a man.

For the last vital cellular requirement mentioned above, maintaining itself in a healthy living state, a cell must manufacture a myriad of chemical products, among them many different kinds of proteins, including enzymes, structural proteins, and sometimes hormones; polysaccharides (complex sugar-type substances which are the "batteries" of living cells); and the so-called secondary metabolites, including such organic chemicals mentioned in Chapter 1 as vitamins, amino acids, and alcohols. Basically, all organisms, whether microbes or humans, are built out of proteins. Enzymes are a type of protein that carry out maintenance and "worker" functions in the body, and hormones are proteins that carry out more complex regulatory functions.

The task of the genetic engineer is to subvert this natural chemical manufacturing process to their own ends in order to turn the cell into a factory that makes commercially useful products. Of major importance are the problems that crop up when the genetic engineer takes over *E. coli*'s processes, because these problems determine how long it will take to commercialize potentially useful products. We'll also consider organisms other than *E. coli* that might have important uses as hosts.

Everything in this first science section is vital to understanding genetic engineering, and everything in the process of DNA function and cellular life that does not relate has been eliminated. The later science sections range a little wider over the field, discussing some of the specific tasks of genetic engineers and some of their latest tools. Finally, the last section (Chapter 9) continues the discussion all the way through manufacture of a protein and explains some of the pitfalls in commercial "gene-splicing." The last science section also includes a short course in the genetic code, a language far simpler than French or German, but one in which the most profound secrets of life are written.

DNA THROUGH THE
SCHEMATIC LOOKING GLASS

DNA is a chain molecule containing millions of "links," but even such a long molecule is too tiny to be made visible except with the most modern techniques and the most powerful electron microscopes (see the photograph below). To offer some idea of the incredible proportions involved, an average bacterial DNA molecule is about one twenty-fifth of an inch long, and that certainly is *long* enough to be visible; but its *width* is only one millionth of that. By comparison, if you imagine that molecule as two intertwined chains whose links are an inch wide, the chains would

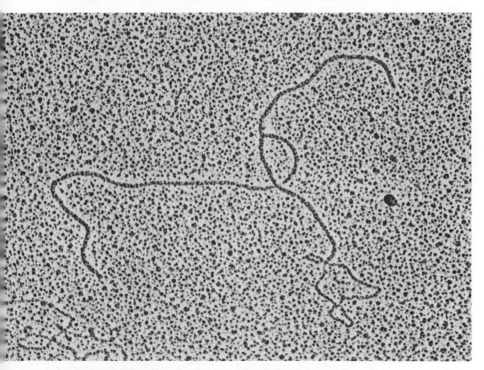

A bacterial DNA molecule magnified 80,000 times.

be more than fifteen miles long. The DNA found in every human cell would be nearly two thousand times that long—nearly long enough to circle the globe.

The puzzle of the structure of DNA was finally cracked by James Watson, an American, and Francis Crick, a Briton, at Cambridge University. That discovery is considered the major milestone in modern molecular biology, and it earned Watson and Crick the Nobel Prize in 1962.

The molecule resembles a ladder that has been twisted to resemble a spiral staircase. To visualize this all-important structure, look at it under the "schematic microscope" we've created in Figure 1. The schematic microscope will allow us to see biological life at various levels of magnification in artist's renditions. Each of the DNA representations shown is used by scientists to highlight different sets of qualities. At the left of Figure 1a is the double helix, closest to DNA's natural configuration. At the right, a major activity is shown: the unwinding or unzipping of the two strands, the first step in DNA's replication. Of all DNA's characteristics, none is more amazing than this ability of a lifeless chemical molecule to reproduce itself, as we'll witness in a moment, and that ability is directly related to its structure. The unwinding of the two DNA strands can occur because the two "backbones" you can see as wavy lines on the double helix are very strong; all of the identical backbone elements are held together by strong molecular bonds, just as our vertebrae are strongly joined yet supple. The single steps in this spiral staircase are actually formed from two separate pieces, joined edge to edge. Each separate piece is called a *base* (solid vertical lines in the diagram). These base pairs are *weakly* joined, and when the DNA unzips, the steps split up the middle while the backbone remains intact. Figure 1a shows the strong bonds as solid lines, the weak bonds as dotted lines.

Figure 1b shows the zipped-up molecule in a different schematic, appropriately called the ladder representation, in which the helix has had the twists taken out, and Figure 1c shows the ladder at higher magnification to reveal some of the substructure. Here you can see the most important aspect of DNA from a genetic or hereditary point of view. The letters represent the bases, which are chemical compounds: adenine, thymine, cytosine, and guanine. *The order of these bases along the DNA backbone carries all the information of heredity.* The bases A, T, G, and C can be thought

a

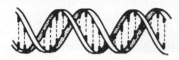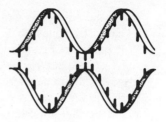

b

c

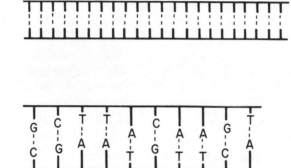

Figure 1. Three schematic representations of DNA.

of as composing the four-letter alphabet previously mentioned. The order of the bases pictured in the lower strand of Figure 1c— CGAATGTTCA—then can be seen as a written instruction in DNA language that will eventually result in the ordering of amino acids into various proteins. If you were to change the order of the bases, you would change the instruction. In a living cell, DNA language is read not by eyes but by other molecules, which then carry out its instructions. Later, we'll learn how the cell reads and interprets DNA language. To get some idea of how significant these commands are, consider that the substitution of just one "letter" for another in a single command out of millions in the human DNA chain is the cause of several genetic diseases, including sickle-cell anemia.

At this magnification the extremely important principle of complementarity can be seen. The G base (guanine) on one backbone *always* bonds to a C base on the other backbone to form one base pair, and an A base (adenine) always bonds to a T base (thymine) to form the second type of base pair. One of Watson and Crick's major realizations leading to the unraveling of the DNA puzzle was that although the shapes of the four bases are very different from one another, the GC base pair is almost perfectly congruent with the AT pair. This is shown in the schematic by the equal length of the "rungs" between the two DNA strands.

Looking at the magnified strand, complementarity tells us that whenever there is a T base on one strand, there will be an A base opposite. That means that the two strands are not identical, nor are they simply in reverse order from one another. Each is a complement of the other: if one strand reads . . .CGAAT. . . , then the complement will read . . .GCTTA. . . as the schematic shows. Further, knowing one strand's base sequence, we can always predict the other's. The chromosomal DNA of the bacterium *E. coli* is several million bases, or letters, long, so it encodes plenty of information on how to make the cell and carry out all of its life processes.

A MOLECULE THAT REPRODUCES

Chemical molecules form when enough atoms of the right kind are present in the right quantities and conditions, but we know of no other molecule that itself becomes a template, or pattern, for the creation of two identical "daughter" molecules, an analog and vital precondition for the reproduction of a cell into two identical daughter cells. The process we're going to oversimplify schematically has been developing and evolving for millions and perhaps billions of years.

After DNA unzips, the replication into two identical daughter molecules occurs. Simply put for now, "worker" molecules (enzymes) in the cell attach the correct base-plus-backbone fragment to each of the *two* unzipped halves, creating a whole new double helix out of each single parent strand.

It's easy to see how complementarity is used in DNA's replication into two daughter molecules. Figure 2a shows a small piece of double-stranded DNA, and Figure 2b shows the strands partially unwound. From a pool of bases (derived from food the

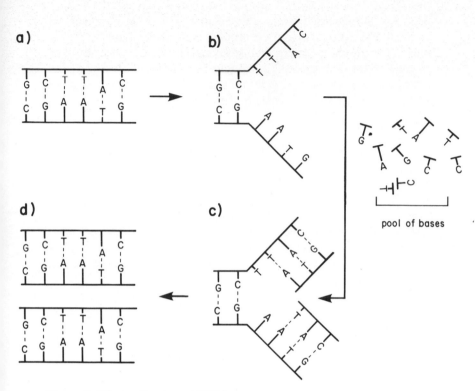

Figure 2. The replication of DNA.

cells eats), the worker enzymes attach the correct complements shown in Figure 2c, where a partially replicated molecule is pictured. In Figure 2d, replication is complete—there is one daughter DNA molecule for each of the two daughter cells that will be built. A molecular biologist will recognize that the replication has been so radically simplified that it is not scientifically correct, but it does show how the key principle of complementarity is involved in DNA replication.

If so much is known about the duplication of the vital DNA in cell replication, it might seem an easy jump to see how the rest of a living cell is duplicated, but that turns out to be a difficult, unsolved puzzle. Molecular biologists know that to a large extent, many cell components self-assemble. This remarkable property of many biological structures can be likened to shaking a box of Lincoln Logs and tossing out completed log cabins. We have to fall back on our old answer: that self-assembly must have evolved through millions of years to its present sophistication, just as DNA, enzymes, and all the molecules we talk about are the latest steps in that same evolutionary development.

HOW *E. COLI* MAKES ITS LIVING

E. coli has several qualities that long ago recommended it to biologists for study. Although a dweller in the human intestine, it is harmless to us; it reproduces rapidly, and it can be cultured in the lab on a diet of sugar water and a few other simple, inexpensive nutrients. These virtues led to its becoming the best-known microorganism in molecular biology, and for this reason it was the first of the genetic engineer's cellular factories.

A cell's single most vital job is to reproduce, and at that *E. coli* is a champ. One cell divides into two daughter cells identical to the original every twenty minutes, under optimum conditions. That means we get four cells after forty minutes, eight cells in one hour, sixty-four in two hours and, in a mere ten hours, a teeming colony of over one billion, each identical to the original parent (in the absence of mutation). The great power of genetic engineering is tied to this wholly natural reproductive ability; if just one cell is built right, the production boom almost takes care of itself.

E. coli is unicellular, each cell living alone in its environment with no attachment to others of its type. Humans, of course, are multicellular: billions of cells of such types as skin, muscle, and blood are organized into a community, forming a (usually) functional whole. "Microorganism" means too small to see without a microscope; one *E. coli* cell is about 1/25,000th of an inch long and a fourth as wide. An *E. coli* cell and some of its internal components are shown under the "schematic microscope" of Figure 3. To see how the components interact to build, maintain, and reproduce the cell, consider this fable of one developer's awesome success in creating a planned community.

Let's call our protagonist Frank Lloyd Levitt, a home-builder with a terrific scheme for building whole subdivisions of identical houses with which he dreams of covering the suburban New York

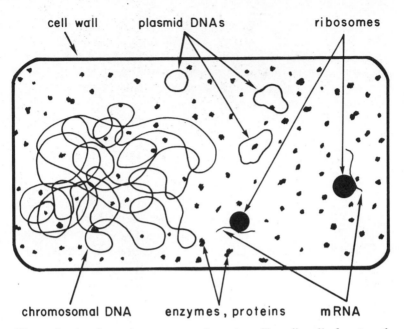

Figure 3. A schematic representation of an E. coli *cell showing the major components involved in protein manufacture.*

countryside. The key to Levitt's method is his "pilot" house; within it is the blueprint for the whole subdivision, explaining step by step what materials are needed and how and when they are to be joined. First, supplies such as block, lumber, and nails have to be delivered to the construction site. He also needs a source of energy for construction. And once a house is built, he'll need whole new categories of equipment—brooms, cleansers, paint— to keep the place livable. An *E. coli* cell would similarly need to bring in food as a source of both raw materials and energy for construction, and later would need enzymes to "keep house."

The pilot blueprint, of course, is the most important single item, and Levitt doesn't want to carry it around and risk damaging it, so he photocopies various sections of the blueprint as needed. DNA is the cell's pilot blueprint, and the cell also has a copying machine, called RNA polymerase. The "copy" of a given DNA segment this copier makes is called messenger RNA (mRNA); it's pictured in Figure 3 as a shorter curvy line than the "template" DNA.

But there is more to builder Levitt's design than the blueprint; there is also an automated workbench. Mr. L simply feeds a copy of one of these blueprint sections into the workbench; then electronic feelers scan the segment and robot arms grab the wood, bricks, and mortar. Whatever is called for in the on-site blueprint segment (mRNA) is assembled on the workbench automatically.

E. coli's robot workbench is composed of large, nearly spherical structures called ribosomes (pictured), and smaller molecules not shown, including enzymes. The ribosomes grab onto the mRNA blueprint copy. Through subtle molecular interactions, they read and execute the instructions on the mRNA blueprint at this cellular workbench. They hook protein building blocks called amino acids into chains of different sequences (much as each line of this page is unique as a chain of different sequences) to form the cell's structural proteins and the proteins and enzymes that will carry out "housekeeping" functions. (How this translation of DNA and mRNA language into the language of proteins is accomplished is detailed in the last science section.) A protein is generally nothing more than a chain of amino acids, folded and wound into a three-dimensional glob.

At this point, Mr. L is on his way to becoming the most successful real estate developer of all time. From one little matchbox house, his blueprint-reading robot workbench first duplicated the pilot house—right down to the new master blueprint and new robot workbenches inside to build yet more houses. The duplicates went on to make duplicates, each duplicate containing new benches and blueprints.

Now, twenty-four hours later, he has laid out East Coli, a subdivision of ten billion homes. He's squeezed out the competition on Long Island and spread through New Jersey. Nothing, he feels, can go wrong. But as is so often the case in this dog-eat-dog world, Mr. L has grown inattentive. He's so busy expanding that he failed to notice his scheming nephew, the genetic engineer Dr. K.

Invited to watch the amazing self-replicating subdivision, Dr. K notices that the automated workbench will build anything it reads on the blueprint. A bountiful supply of raw materials is on hand, and Dr. K realizes that if he stitches his own little blueprint segment into the master, then every house built will have an expensive sports car in the living room. All Dr. K has to do is to break out a wall, drive the car off, and he'll be in the car business without lifting a finger. By the time Mr. L finds out, Dr. K will have a fortune in once-rare cars. Among the problems: if the robot workbench uses up too much material and energy building a sports car, then it won't be able to build more houses, and that in turn will limit the number of sports cars that can be built.

In a nutshell, that is the task and one of the major problems facing the genetic engineer. *E. coli*'s ribosomes (cellular particles composed of RNA and protein that function as the site of protein synthesis) will build exactly what is put in front of them, following instructions coded on the mRNA, regardless of the origin of that mRNA. Genetic engineers have to supply the ribosomes with a good supply of the mRNA for their desired proteins, no easy task. They must make equally sure that in addition to making a lot of the particular mRNA for the desired proteins, the cell continues to make enough of its own proteins' mRNA for it to survive and reproduce.

Today's "flashy sports car" of genetic engineering is interferon, and large quantities of various types have already been produced in *E. coli*. The bacteria have no use for these animal proteins, and although we do, our cells make them in extraordinarily tiny quantities. Still, interferons are made on human-cell ribosomes just as they can be made on *E. coli*'s.

As we mentioned earlier, a gene is nothing more than a region of DNA. *Generally we think of the gene as that region which carries instructions for a single protein.* Thus, one gene codes for the production of one kind of interferon protein. Once the genetic engineer has slipped the interferon gene into *E. coli*, if he or she inserts special signals telling the cell to make many mRNA copies of that gene, the end result is a high rate of interferon production. We'll see that these signals exist and are very important to genetic engineering.

Now we're really in business: we've engineered each cell to mass-produce interferons, and hitched that engineering feat to the cell's native ability to mass-produce itself. That's why genetic engineering offers such potential for scaling up production of rare and now-costly drugs and chemicals.

Another major tool of the trade is plasmid DNA, seen as circular DNA in the schematic microscope (Fig. 3). For its function, let's head back to the subdivision of East Coli. Despite the greed of his nephew, Mr. L's subdivisions are doing fabulously in the East, and he decides to hit the sunbelt, someplace like Arizona. But he doesn't want to build exactly the same house in Phoenix that he's building on Long Island (though many have tried). The basic blueprint is still fine, but it must be modified to take local conditions into account. Central air conditioning will be a must, for example. What he needs is something like his nephew's intrusive commands: supplemental blueprint units that will vary from one locale to another, this one adapting the houses to dry heat, that one to extreme dampness, still another to months-long cold and deep snow.

These supplemental blueprints we've envisioned are surprisingly similar to the little circles of plasmid DNA contained in many bacteria. In contrast, chromosomal DNA—the master blueprint—contains all the information necessary to survival under "normal" circumstances. But *E. coli*, an extraordinarily well-traveled organism, might find itself in an inhospitable locale—for example, an environment containing penicillin or another antibiotic. A normal *E. coli* would die in the presence of penicillin, while one with a plasmid containing instructions for defense would survive.

Where does the plasmid come from? In a population of *E. coli* cells only a very few might contain the plasmid for resistance to a particular antibiotic. In function, the plasmid DNA would instruct the cell to make a protein that would combat the antibiotic. If you now introduce the antibiotic into the colony, all the *E. coli* without the plasmid will die. That means the future colony will be composed only of cells containing the plasmid, because they are descended from those survivors that had it. But surprisingly, cells can also pass plasmids to other cells that don't have them via a very complex mechanism. That means that when an antibiotic is introduced into a colony, some cells without plasmids will get them and themselves survive, in addition to passing the plasmid on to their descendents. Plasmids are duplicated as their hosts divide, and that is of vital importance to their use in the gene-splicing tool kit.

The genetic engineer rebuilds plasmids outside the cell, inserting into them the gene to be "expressed" (that is, made into its corresponding protein product). The engineer then places the plasmid in a broth containing "host" bacterial cells, and the hosts incorporate the plasmids into their own cells and pass the information from generation to generation as if picked up naturally.

Everything discussed so far can be summarized in two simple diagrams and as many sentences. The diagrams are standard scientific shorthand for chemical reactions and other types of transformations, showing the agent bringing the reaction about above the arrows, and demonstrating the general flow of instructions for a living cell and for our homebuilder.

$$\text{DNA} \xrightarrow{\text{(RNA polymerase)}} \text{mRNA} \xrightarrow{\text{(ribosomes)}} \text{protein}$$

$$\text{Master blueprint} \xrightarrow{\text{(copy machine)}} \text{copy of blue-print section} \xrightarrow{\text{(robot workbench)}} \text{house}$$

The genetic engineer wants to speed up any or all of those steps, and the snowballing effect of increased rates on one another is what is called "overproduction" or "maximization of gene expression." But before getting too far afield on methods, we need to look more closely at the subcellular pieces that play key roles in production or are desired proteins themselves, and to start with, we must take a closer look at DNA itself.

CELLS: THE INSIDE STORY
Proteins, Secondary Metabolites, and Other Cellular Goods

Some molecules such as oxygen, table salt, and water are small clusters of the same or different atoms. But most of the molecules of interest in biology are long chains of subunits, and even the subunits themselves may be fairly complex clusters of atoms. Such long-chain molecules are called polymers, and all living things contain two important classes of them: nucleic acids (DNA and RNA), the instructions for creating and maintaining life; and proteins (also called polypeptides), which we can think of as the "stuff" of life. Let's start with the nucleic acids.

As in Figure 1, each of DNA's two chains is composed of the

bases A, G, T, and C. Therefore we can represent a section of chain simply as letters and dashes as in this hypothetical sequence:

......G-G-T-C-A-A-T-G-C-T......

In this way, DNA resembles a four-colored necklace, with each base a different-colored bead.

Proteins, too, are necklacelike chains, but their "beads" are combinations of twenty different chemicals called amino acids. The names of all the amino acids and their abbreviations are given in Table 1. Thus, a piece of a protein molecule made of a chain of the amino acids methionine, glycine, alanine, and valine can be written schematically this way:

........Met-Gly-Ala-Val.............

Table 1. The Amino Acids

The 20 Amino Acids	Abbreviations
Alanine	Ala
Arginine	Arg
Asparagine	Asn
Aspartic Acid	Asp
Cysteine	Cys
Glutamic Acid	Glu
Glutamine	Gln
Glycine	Gly
Histidine	His
Isoleucine	Ile
Leucine	Leu
Lysine	Lys
Methionine	Met
Phenylalanine	Phe
Proline	Pro
Serine	Ser
Threonine	Thr
Tryptophan	Trp
Tyrosine	Tyr
Valine	Val

The sequence in which the twenty amino acids are strung together determines the protein, just as the sequence in which the four nucleic acid bases are hooked together determines the instructions of the DNA blueprint. The analogy to written language is obvious. DNA language consists of a four-letter alphabet. Similarly, protein is a language written in a different, twenty-letter alphabet. The analogy is not just a learning device. So close are the parallels with language and its uses that molecular biologists describe the flow of genetic information in a cell this way:

$$\text{DNA} \xrightarrow[]{\text{(transcription)}} \text{mRNA} \xrightarrow[]{\text{(translation)}} \text{protein}$$

The making of an mRNA copy from DNA is called transcription because it is virtually the same as transcribing notes or comments from one medium to another without altering their form.

Making a protein from the instructions contained in mRNA is called translation because it represents taking instructions written in one language (A-G-T-C-etc . . .) and translating them into a useful form in another language (e.g., Met-Gly-Ala-Val).

We talk about the details of transcription and translation later, but these processes carried out inside the cell are the same except for details, whether the cell be *E. coli*'s or your own.

Although mRNA is a good copy of DNA, there is one difference that occurs in transcription. In place of the base thymine (T) in DNA, molecules of RNA contain a similar base called uracil (U). The difference is more in the name than the shape, for although an mRNA sequence might read -U-U-U-, its informational content is the same as DNA's -T-T-T-. Why nature chose to use this similar but different "letter" only in RNA is not fully known.

Proteins in turn can generally be considered in two classes: structural and housekeeping. "Structure" is as simple to understand as bricks, wood, or mortar, but the housekeeping functions of a protein are more complex and involve the actions vital to living: digesting food, breaking it down to cellular building blocks and/or deriving energy from it, clearing out waste, and so on. While structural proteins may have some commercial value, the housekeeping proteins are far more interesting in that respect; they are the enzymes that carry out all the cell's chemical reactions, manufacturing a tree's wood, the vitamins a human needs, or the antibiotics produced by some bacteria. These are not one-step operations, however. The enzymes or vitamins or antibiotics, produced by bacteria, operate in often-long series of chemical reactions involving perhaps dozens of different enzymes, each doing a single reaction job in its turn. Biochemists refer to these chemical reaction sequences in the cell as "metabolic pathways."

In Chapter 1 we mentioned that enzymes are good catalysts because they are so specific. Most chemical reactions involve unwanted by-products, but one enzyme will carry out only one chemical reaction to yield a single product. That means we don't have to separate the desired product from other residues produced along the way; we can eliminate the usually costly steps of purification. We mentioned previously that industrial force, through blast-furnace temperature and high pressure—major contributors to environmental and energy problems—can be replaced by enzymatic power. Enzymes have to work under mild conditions because they evolved working inside living organisms.

Wood or scrap conversion to alcohol depends on the fact that some fungi and other organisms use wood as food. Most organisms can't break down the cellulose-rich wood into glucose because they lack the enzymes, called cellulases—a shame because, strong as it is, cellulose is nothing more than a long chain (polymer) of glucose molecules. If *E. coli* could be genetically engineered to make large quantities of these wood-digesting cellulases, large quantities of glucose could be made from waste wood, scrap paper, and such farm waste as corn husks. The glucose could then serve as a food source for yeast—in yet another conversion chain—as the yeast turns glucose into alcohol as fuel or feedstock.

An enzyme-catalyzed reaction is taking place under the microscope in Figure 4. Figure 4a shows a simple chemical reaction, a *degradative* one in which a complex chemical A-B is to be broken down into parts A and B. Enzymes actually carry out both these degradative and *synthetic* reactions which, as the name implies, involve building more-complex chemical structures out of smpler ones.

In this degradative reaction, the catalytic enzyme fits its shape to the A-B molecule complex and exerts attractive forces on it, breaking the bond between the A and B parts, thus releasing the two separate molecules. The enzyme is then free to repeat the process on another A-B complex.

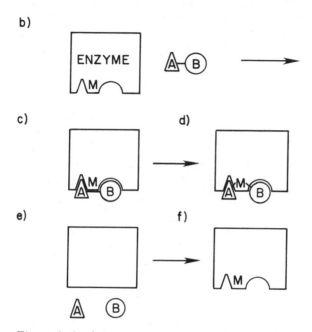

a) enzyme catalyzed reaction:

Figure 4. A schematic representation of enzyme function.

Figure 4b shows the catalyzing enzyme. Its most noticeable feature is its surface, which contains two pockets corresponding in shape to the A and B groups—such "lock and key" surface complements are common in enzymes. The two chemicals catalyzed by the enzyme are known as substrates. The enzyme's specificity is ensured both by the pocket shape and by varying, often-subtle attractions between chemical groups. Real molecules have incredibly complex and varied shapes, and enzymes have evolved to mirror those shapes. The actual specificity of enzymes is far more complex than we've depicted.

Figure 4c shows the enzyme bound to the substrate. Usually the attractive interactions between enzyme and substrate are weak compared to the strong chemical bonds that hold the substrate together, and that's where things get interesting. How does an enzyme molecule break this strong bond? Think of the chemical bond as the force holding two magnets together, for chemical bonds are attractive forces, as are magnetic forces, though certainly not the same. You might unstick the magnets by yanking them apart. But an alternative would be to bring in an electromagnet; if even momentarily the electromagnet's attraction for each magnet is stronger than their pull on each other, the bond is broken. Look at the enzyme area denoted "M." If the attractive forces in this region alone are strong enough, they can help break the strong bond between A and B, as shown in Figure 4d of the figure.

But now the enzyme must release the broken pieces of A and B into the medium, and this can occur in several ways. Since the *overall* bond between substrates and enzyme is weak, the A and B molecules might just jiggle off through natural vibrations. If the overall bond is too strong for that, the enzyme might undergo a slight change of shape so its "lock" pockets no longer fit the "key" substrates. Figure 4e shows an enzyme "spitting out" A and B by flattening out its pockets, only to revert to its most stable form in Figure 4f—ready to go again.

Enzymology is among the most complex fields of biochemistry, and chaining enzymes to our own industrial production is not a simple task. We explore some interesting properties of enzymes and some of the complexities of metabolic pathways in a later science section.

chapter 3

Genes and Genies

Tom Roberts, bent over a small plastic vial, releases a thin rubber tube from his lips and a stream of clear liquid flows from the attached pipette into the test tube. The quiet laboratory room is not new or particulary modern-looking—black stone bench tops, chart on the wall, large jars with small plastic test tubes with a refrigerator to store them. Another man works nearby, equally silent.

Working from cookbooklike formulations in his head and out of scientific papers, many of which he helped write, the soft-spoken, self-effacing Roberts is performing an engineering feat of dazzling complexity and precision. He is altering the genetic makeup of a colony of bacteria so that as each cell carries out the building and maintenance functions dictated by its genetic makeup, it will also make proteins alien to its heredity. All the noise and hype of Wall Street focus on quiet rooms like this. But then, most fundamental discoveries in science, however enormous their eventual consequences, take place in such rooms. Not surprisingly then, the quiet here belies a real excitement, a feeling of competition at high energy, and sometimes you get a real whiff of powder.

A different scene: a Wall Street stock analyst who prides herself on critically appraising the new recombinant DNA firms, argues with a scientist over whether the gene-splicing techniques are ready for the marketplace. This is perhaps the most critical question in the efforts of business and science to understand one another.

"I've heard people boast that their scientists had cloned a protein," the analyst says—meaning that actual production of the desired end product had been achieved. "But when you talk to the scientists themselves, they're very straightforward. They tell you, 'Cloned it? No, we've just inserted the gene for it.' "

From a practical *or* scientific viewpoint, snipping and inserting new genes into bacteria is relatively simple compared to getting the host bacteria to "read" the new gene correctly and carry out its instructions.

The scientist rejoins: "Listen, Tom Roberts can get *E. coli* to spit between 0.5 percent and 5 percent of its proteins as the desired protein. He's one of the top gene-splicers. And he can do that almost routinely: get 'expression' of the gene as protein, and also get high yields. He's just one of the best there is."

Soon after that conversation, Roberts raised that percentage from 5 to nearly 15. Now, achieving such overexpression percentages in *E. coli* is routine in many better laboratories, and achievement levels in *Bacillus subtilis* bacteria and *yeast* are at the exciting "frontier" where *E. coli* experiments were two years ago.

The lingo may be new on the street, but an old feeling of challenge comes through: this is Harvard Medical School where Roberts works, one of *the* mountain peaks in the scientific range. Harvard, M.I.T., Cal Tech, Stanford, UC San Francisco, Cold Spring Harbor Laboratories on Long Island—these and a few other institutions define the big time in genetic engineering. If something is worth doing at all here, it's done well, then better, and always fighting for best.

Roberts is one of the acknowledged experts in "overproduction," the newest set of techniques for getting very high quantities of proteins such as insulin or interferon. He was part of the Harvard team under Dr. Mark Ptashne in 1978 that combined a variety of chemical command signals into *E. coli*'s genetic material to force yields in quantities staggeringly larger than had previously been obtainable. For example, of a particular protein which in nature would be cranked out at the rate of 100 molecules per cell, Ptashne's group was able to get 200,000 molecules per cell. Roberts says that the yield today is 600,000 molecules per cell.

No one has to tell someone in business why yield is important. If your bacterium makes ten times as much insulin as your competitors' at the same cost and quantity, you win and they lose. But to scientists, yields mean something entirely different. They spell the difference between interferon's being an arcane protein whose structure cannot even be guessed and a substance common enough for researchers to study. And with this and other substances, radical yield improvement can transform a laboratory curiosity into a bulk drug or chemical.

The industry springing up around genetic engineering is going to take much of its character from these molecular biologists, who, as a group, have a distinct personality. They are single-minded high achievers, capable of great intellectual focus. The top among them are Phi Beta Kappa and/or honors graduates of leading colleges and universities, but they stand out among other kinds of scientists as intensely visual, as imaginative rather than analytical.

Dr. David Baltimore, awarded the Nobel Prize in 1975 for discovering a crucial DNA enzyme, recalled in a recent interview his first interest in biology, sparked in undergraduate honors seminars at Swarthmore College, which had no laboratory. "I wanted to *see* something," he said, a comment wholly in character for this group of scientists. "I could tell that biology was moving into a revolution and the faculty could not. I'd read all the books and papers, but the impetus to see something got me on the trail I've been on ever since."

That trail led the young Baltimore to Cold Spring Harbor, the famous biological research laboratories on Long Island's North Shore, where his requests to see some bacterial viruses led to acquaintance with those who would be his mentors and later his colleagues at M.I.T. That trail would broaden into cancer research, immunology, and the joint founding of Collaborative Genetics, one of the early gene-splicing firms, in Waltham, Massachusetts.

James Watson, mentioned earlier as co-winner of the Nobel Prize for first elucidating the structure of DNA, claimed in his memoir *The Double Helix*[1] that he didn't know any biochemistry when he began trying to solve the toughest chemical riddle in biochemistry and genetics. That is probably an exaggeration, but a former Harvard colleague says, "There may be a germ of truth to it. He comes across as exceptionally bright, and he probably has a tremendous *visual* imagination."

Molecular biologists are also practical intellectuals. When they needed a tool to supplement the chemist's standard ultracentrifuge, a high-technology piece of equipment that can cost multiples of $10,000, they devised a system known as gel electrophoresis that makes use of basic laws of motion and molecular weight. The cost of the little plastic gizmo that carries it out: about $30.

Baltimore says of Watson: "I was just talking to him about the lab, and it occurred to me that it's part of his genius—and it is genius—to be able to put together a laboratory to successfully carry out advanced biological research; to put together the right people and the physical facilities, and to see the direction things must go, to understand the dynamics of intense, extremely bright people working together."

Baltimore contrasts his colleagues with scientists in other fields this way: "People in biology always had to be somewhat better grounded in reality. We lack theoreticians . . . largely because it is so empirically based a field. You must maintain an involvement over a very long time to be good." And that is why, compared to theoretical physics, he says, "There are very few young people at the top. It takes time to come to grips with all the procedures and techniques and to have them mesh with the theoretical."

Our popular image of scientists at leisure was left us by Einstein: playing classical violin, reading, abhorring physical exercise. But many biochemists are "physical," the kind of people who blow off steam hard rather than easy. When Tom Roberts was Lynn Klotz's graduate student, they played every pickup basketball game they could find, along with others who are still their major colleagues and contacts. Mark Ptashne is known to be a good squash player and top-level Ping-Pong player—as well as a devoted violinist—and Jan Pero is said to be an excellent volleyball player. Vicki Sato describes herself as "nonathletic"—but her major avocation is ballet, and when she resumed dancing as an adult after years of absence, "I danced for hours every day"—as demanding, intense, and physical an activity as any sport.

And these examples are not anomalous. A top yeast geneticist, Gerry Fink, was a varsity basketball player in college. Herbert Boyer, the founder of Genentech whose teaming with Stanley Cohen turned laboratory gene-splicing into an industry, played football in the Pennsylvania country where it is king. He was voted best athlete by his high school class, but foresaw a career in molecular biology when he did a college report on DNA soon after Watson and Francis Crick unraveled its secrets.

Of course, such characterizations cannot be universally applied. Dr. Walter Gilbert, Nobelist and a founder of Biogen, doesn't think of biologists as more or less physical or practical than other scientists. In fact, he began his career in theoretical physics, which he taught at Harvard for several years before switching to molecular biology. Dr. Shing Chang, a leading expert on the bacterium *B. [Bacillus] subtilis*, also feels that such statements overgeneralize.

Significantly, women make up perhaps half the top researchers in genetic engineering fields, a marked difference from the situation in most other hard sciences and engineering. An officer of one rDNA firm said half its program managers and other doctorate-level staff members are women, as are about half its bachelor-level staff.

Sato said, "It's hard to tell whether women are now more conspicuous just because they've been in biological sciences longer or if there's some other reason that biology was always less sexist" than other sciences. But she added that her women colleagues in physics and other Harvard departments faced "a great deal more sexist pressure" than those in the biological sciences.

This group on the cutting edge of molecular biology is small and ingrown. Relatively few men and women have mastered the science of gene-splicing or its related skills, and most of them know each other. They studied with and under one another, competed in adjacent laboratories, collaborated as postdoctoral students. Some date one another, a few are married to each other.

It's been suggested that gene-splicing is really simple, that anyone can do it. But Princeton University's Dr. Jacques Fresco points out, "If all you want is to put some kind of genetic information into a foreign cell but you don't care what—that is, you want to clone something—then, sure. I could devise an experiment for undergraduates and most would succeed at it. But if you're talking about producing a *desired* protein in *quantity*, then there are only a handful of people in the country who are masters. A handful? Maybe twenty, maybe fifty—not a large number."

With the rapid progress in the field, that statement already is outdated—as we noted, many good labs can now achieve this level of overproduction. But its essence remains true: those on the *new* cutting edge, carrying out the analogous experiments that will extend the frontier farther, are few in number and known to one another.

Molecular biologists are among the most intensely competitive people in science. Some executives have questioned whether academic scientists will be able to take pressure of industrial competition, to slug it out in the "real world." There are a lot of "real world" questions the scientists worry over—but whether they can compete is not among them. Asked about rivalry at Harvard, Vicki Sato said, "The competition [in science] is murderous. People right down the hall will be working on the same thing you are, and you're trying to beat one another. As soon as one of you publishes, the other hops on—to find a mistake, or to use the results if there are no mistakes. Competition is the name of the game, for results, for grants, for the best academic jobs, and for tenure."

In Horace Freedland Judson's thorough history of the genetic revolution, *The Eighth Day of Creation*,[2] Ptashne recalled months of "absolutely hair-raising" competition with Walter Gilbert, who was working just a few labs away at Harvard. The two were trying to find the "repressor," a chemical that prevents genes from ordering the manufacture of proteins. Ptashne recalled his exhilaration and terror, a feeling of certainty that one of them would be the discoverer, not knowing which of them it would be.

In 1967, each found a different repressor. They both won the race, though Gilbert had a slight edge. The repressor turned out to be of vital importance to molecular biology and to its offspring, genetic engineering.

Gilbert now says the "competitive element between Mark and me has been overstated." But when asked if scientists are not in fact used to intense competition, he laughs and says, "I don't think that will be a problem" as scientists enter the high tech business. He argues that since the scientists are *more* competitive than businessmen they won't have any trouble adjusting to the career change. When interviewed early in 1982, Gilbert had just begun a year's leave of absence from Harvard to devote himself to Biogen, where he was chairman of both the board of directors and the science advisory board.

Gilbert resigned his tenured professorship before the end of last year to hold the reins of Biogen full time. Because Biogen's headquarters are in Switzerland, "he's become kind of a jet commuter" between there and Cambridge, according to Sato, who had just had dinner with him and his family.

Still, there are differences in the competition experienced in a university and that in private industry. The race in academic science is to learn new information and get it published as fast as possible. The rush in industry is to produce the most·salable product, and that may require keeping results secret until the moment is right. That difference obviously can have major effects on the speed with which information flows and thus on the overall rate of development of new products.

But in one respect, life may be easier in business. In science, the first person to publish almost always gets all the credit, even if the second is not far behind and has done a better job. But in industry, you don't always have to come out with your product first. No one remembers slight time differences. If your product is better, or cheaper, or you market it more skillfully, you can come out very much ahead without being first.

THE RACE

The genetic revolution began in 1953, when biologist Watson and physicist-turned-biologist Crick, both at Cambridge, got their hands on the best set of X-ray crystallographs of DNA available, taken by Rosalind Franklin. With Crick's knowledge of physics and chemistry and Watson's visual intuitions and "hard thinking" (a colleague's words), they built the first model of deoxyribonucleic acid. The model climaxed an intense race with Cal Tech's Linus Pauling, then considered the best chemist in the world, and others in both Europe and America. It was a race full of clashes of wills, of whimsy, and of bitterness.[3]

Watson sharply criticized Franklin in his memoir for pushing off in wrong directions, even though he would not likely have found the secret without her X-rays, and Watson drew Crick's scorn for publishing *The Double Helix*, which his erstwhile partner found undignified and too critical of colleagues. Maurice Wilkins, who shared the Nobel Prize with the two, was also sharply critical of the late Franklin, but he said to science historian Judson: "DNA, you know is, Midas' gold. Everyone who touches it goes mad."[4]

If not truly the cause of madness, DNA is surely the driving inspiration of those pursuing the most intimate secrets of life and its controls—a molecule described by more than one scientist as elegant: spare in design and symmetrical, like a Greek temple or a perfect circle.

Watson and Crick have remained at the forefront of the science they spurred, Crick a leader in the following decades in helping crack the genetic code and studying its molecular details. Watson has run Cold Spring Harbor, which is now involved in major genetic engineering undertaken in large part under a $7.5 million, five-year contract with Exxon.[5]

Scientists spent the 1960s unraveling the tortuously complex details of how genes "express" themselves to order the manufacture of proteins, then how the strings of molecular beads called amino acids fold themselves into often relatively massive three-dimensional proteins.

To realize how critical the relation between DNA-function and protein-function is, consider the case of sickle-cell anemia, a disease that primarily affects blacks, and kills by reducing or eliminating the ability of the blood to carry oxygen. A *single* incorrect instruction in the DNA chain puts the wrong amino acid in just one link of this massive protein chain. This distorts the molecule's shape just enough to prevent a critical group of iron atoms from attaching to oxygen to carry it through the blood.

If scientists could some day correct that false order in, say, the fertilized egg, the child would never get the disease and might not pass the trait on to his or her children.

Molecular biologists of the 1960s also were scrambling to find the mechanisms that turn genes off and on—the repressors, thought to act like a mask or hand that covers a light until the right chemical conditions remove it and the light can begin signaling.

Galactase is an enzyme that breaks down milk sugar (lactose) in digestion. The gene that directs the manufacture of galactase out of the building blocks available in the cell is normally masked by the "lac repressor." When a milk sugar molecule enters the cell, the lac repressor combines with *it*, moving away from its normal position on top of the gene, which then signals for assembly of galactase to break down the sugar.

Harvard's Gilbert discovered that lac repressor. Perutz told Judson in 1968, "The greatest discovery in molecular biology in

1967, and the one that's received the least publicity, is the genetic repressor—isolated by Walter Gilbert and Benno Muller-Hill at Harvard."

Gilbert won the Nobel Prize in 1980—not for the lac repressor but for equally pioneering work in developing techniques for reading the sequences along DNA molecules—ironically, because a colleague notes, "Had Gilbert won it for the repressor, Ptashne might have shared it too."

This colleague describes Gilbert this way: "He is a brilliant, first-rate scientist, with an uncanny ability to know exactly what is going on in his lab and to separate good procedure from bad. At times arrogant, I feel, but absolutely brilliant."

Arrogance is not a foreign word in big-league science. Ptashne is also described as brilliant, with more than a touch of arrogance and a difficulty in working with people. But Ptashne "was awesome in getting things done," the colleague recalls. "He and Tom Maniatis—then a postdoc [postdoctoral fellow]—set out to work on repressors that I figured would take five postdocs a couple of years. I mean, nice idea, but how were they going to carry it out? Nine months later they'd accomplished what they set out to do. Mark is terrific at visualizing an argument as he's hearing it and being able to sift the good from the bad. A lot of conceptual skills."

David Baltimore, meanwhile, made one of the major discoveries concerning the reproduction of cancer viruses. Many are so simple they do not even have DNA, but possess only a slightly simpler, related carrier of genetic information called RNA. Without DNA, how could a virus take over a cell's metabolism to force creation of more viruses? Baltimore discovered an enzyme that the virus "orders up" that *makes* a copy DNA using the simpler RNA as a template. That was one step in a direction biologists did not believe genetic information could move: it was always supposed to flow from DNA outward, never coming back to alter the DNA.

Baltimore now talks with enthusiasm of the work he is doing in his M.I.T. lab—pure research into cancer. He points out that his 1975 Nobel Prize came for the first experiment he ever did with cancer virus, the first step on what has proved to be a major path for him.

Baltimore and other investigators around the country, for example, now believe that cancer is not always caused by a virus, but is only carried by a virus that has picked up the cancer gene from a cancerous cell. Current concern with discovering cancer genes—several already have been isolated—marks a major change in biologists' thinking. But Baltimore's excitement is over the "pure research," he points out. He does not expect to find a cure for cancer as a result of it. The recent status of cancer-gene research will be discussed at the end of this book.

Still, several years ago, losing none of his desire for pure research, Baltimore began to want to do other things as well: practical things, work that would have an application to problems in the real world. Hence, he developed Collaborative Genetics, a gene-splicing concern.

In research, the early 1970s marked the major milestones in the development of genetic engineering. Scientists knew that each cell in the body contained a molecule of DNA, and that in a given individual, every DNA molecule was virtually identical, repeated billions of times but repeated, presumably, nowhere in the universe outside that individual. How did the same DNA make a blood cell when it was in the blood, and a brain cell when in the brain? How did the single fertilized egg in the womb differentiate into all the tissues and organs?

Stanford's Paul Berg decided that he would look at the DNA of higher organisms to help answer such questions, as we noted earlier, and both "recombinant DNA" and the controversy over risks were born.

The angry, confused atmosphere that sometimes accompanied the debate at public gatherings is believed responsible for at least some major drug companies backing off of recombinant DNA research—enhancing the chances for the smaller gene-splicing firms that dominate the news and, so far, the breakthroughs.

Not long after Berg's 1973 recombinant DNA experiment, Herbert Boyer and Stanley Cohen met. Their aim was no different than Berg's. The difference was in the proposed technique—a technique that became a technology. Cohen had decided that the best vehicle for recombinant DNA would be the plasmid, the circular ring of DNA found only in bacteria. Cohen planned to insert the gene he wanted *E. coli* to use into the plasmid.

In a conversation at a Hawaii convention, Boyer told Cohen he had discovered enzymes that made lighter work of cutting up and rejoining the plasmid ring with its newly inserted gene. These specialized "restriction enzymes" are each designed to cut up very specific DNA sequences—their function in cells is to snip up and thus destroy invading foreign DNA. Another enzyme in Boyer's genetic tool shop would make each end of a cut piece of DNA string attract the other end—in effect, making the ends "sticky" so they could rejoin.

New genes spliced into a ring of DNA were the beginning of recombinant DNA (or rDNA) as a technology, as engineering. A few years later, Boyer joined with venture capitalist Robert Swanson to form Genentech to make such products as insulin and human growth hormone. The race was on to engineer better strains of *E. coli* and learn more about how genes express themselves, and it was at Genentech that the "genies" first discovered "maximization of gene expression"—that is, overproduction.

Meanwhile, back at Harvard, Ptashne's group was running neck and neck with Genentech. Keith Backman conceived many of the key ideas for maximization, Tom Roberts recalls, while Roberts "worked out most of the early technology." Ptashne later would be the major scientist in Harvard's plan to set up its own genetic engineering firm, a plan in which the university would lend its name and laboratory space in return for being a minority share-holder in the company. That plan was dropped after faculty members loudly protested that the university had no business in commercial enterprise. But under Ptashne, the Genetics Institute began as a private company, with heavy venture funding and then the contract work most such firms do in their early years.

Roberts, assistant professor of pathology at Harvard Medical School, has joined BioTechnica International as a consultant. Backman left M.I.T. to work for this new firm full time. Almost everybody is in somebody's company, and the companies are mostly unions of scientists still in academia with small ambitious firms funded by Wall Streeters, drug houses, and giant energy and chemical firms. They fell together, like a chemical molecule, be-cause the elements were there *and* conditions for the particular combination were right.

And suddenly the praise and criticism that passed regularly among those at the top, as it always does, the living and dying by each other's assessments in a small interwoven group, became the rumor and the speculation of Wall Street. Roberts, modest as he describes his work at Dana Farber Cancer Research Center, is praised as one of the country's best genetic engineers, an expert on the molecules that promote gene expression. Asked who *he* thinks is good, Roberts names Genentech's David Goeddel, a major pioneer in overproduction; he's only met him once or twice but knows his work *real* well. "I saw some of his experiments and they were very, very nice."

Top pitchers in the big leagues, and science at these universities is as big as leagues get. Anybody can clone things? Sure, anybody can play baseball, too. But anybody can't play in the World Series, and that's what the craze in the marketplace is about: Who's going to take the Series? Who's going to get that protein spit out so fast and clean that their company and its investors will make millions while the other guys are playing catch-up? Or waiting till next year. Or sinking miserably out of sight.

The sports analogy is helpful in another way in assessing this new, uncertain industry: In high technology firms especially, success or failure depends almost entirely on people. One scientist says, "High tech industries are people-intensive. We don't have a mineral to mine, or a forest to harvest. What we've got is the product of people's minds. How good they are, how well they work together: those are the best things that determine the outcome."

Roberts says Genentech is "a model of one type of company in our field. It *is* its employees, really. You can follow their pattern in this field: if you're interested in a particular [university] lab's process, hire the best postdocs out of the lab for your company." In postdoctoral work, scientists gain valuable research experience before striking out to run their own labs.

DNA Science, the E. F. Hutton creation discussed in Chapter 1, was another type, aiming to back university research in the universities where the best work was being done.

In an interview in July 1981, DNA Science president E. Russell Eggers called the setup "the most interesting cocktail of money in the country," and he was no neophyte enthusiast. Nevertheless, the plan collapsed. Hutton continued to operate a stripped-down version of DNA Science, but without the features that made its original plan unique. It was to have been a marriage of university and Wall Street, and the failure indicates some of the strains of these new partnerships.

But why the straining to create such a partnership? Why all these small companies working for large firms on contract, with other firms in effect trying to broker deals between university labs and commercial gene-splicers? A simpler scenario would have Eli Lilly, Monsanto, Standard Oil, and other giants simply hiring the country's best cloners at any price and accommodating them in corporate labs.

The problem is that major firms have never been able to lure enough top scientists in chemistry and biology away from universities. Those at the "cutting edge" often refer to the work done at many of the drug firms as "third-rate science"—though it may well be first-rate technology and marketing. There are exceptions, but many of the major breakthroughs in pharmaceuticals in recent decades were achieved in hospital and university labs, sold to big drug houses, and there transformed into products which were tested and marketed.

Princeton's Fresco explained, "The prejudice [against working for big companies] is there, warranted or not. The best graduates don't go to work for pharmaceutical houses. Certainly, as an adviser, no professor would suggest it. The best graduates go into academia, into pure research and teaching."

In other words, they are urged toward the cutting edge. A young Princeton professor was overheard talking to an undergraduate who'd just announced she'd decided *not* to go to medical school— even though she was sure she would get in—but to get her Ph.D. and do research instead. "Attaway!" he said, hugging her in rough camaraderie. "I knew you wouldn't sell out."

If medical school is a sellout, where does that leave even the best of the drug houses in the fierce competition for these best and brightest?

Harvard's Roberts said that because of a lack of top scientists, drug companies frequently do not know who is doing what in genetic engineering. "Most of them don't have people who can really talk science, and when they do, it's usually not someone in power."

Suddenly, only the best would do. Where there's a will there's a way. Or maybe this corollary: When science discovers something that has the potential to make a lot of money, and the investment/ industrial community has money burning its pockets, a way will be found to get the money into the science, regardless of the obstacles.

The answer in this case lay in the small companies, beginning with Genentech and the host which have followed or which have adapted their earlier research plans to focus on the new biotechnology. They have emerged, literally weekly, to bridge the gap between university and industrial laboratories. At least, that is the hope. Some small firms were founded by leading molecular biologists, others hired them with major equity shares and, equally important, real authority. So many Nobelists have become involved that their presence on boards of directors and science advisory boards has sparked jokes in both business and science. Science advisory boards are being headed by people who have made the science, and many sit as company directors while remaining at their university research posts. Genentech's Boyer is still a full-time professor at UC San Francisco, for example.

Walter Gilbert pointed out that Biogen began as a group of scientists with venture-capital backing, not the other way around. It was the scientists who recognized that many of their discoveries were ready for development and who knew how that development ought to proceed, he says.

David Baltimore also sees the emergence of the companies as far more than a technological linkup. In fact, he said, long ago he saw Collaborative Genetics as an important part of his work as a scientist, albeit one totally separate from his research. "Things done in basic research have an impact on practical problems, but no one knows what that impact is, it's completely unpredictable. Naturally, you have the desire to utilize research skills in solving practical problems, in bringing real change to life. You've got to somehow mesh basic research and application, so you have to face the needs." The question is, how?

"You *could* do it in a university, but you would scramble for grants that shouldn't be there for work that is going to be profitable. The nonprofit sector doesn't support such work in a capitalist society. When you're working on something with those practical, commercial applications, then you ought to operate as a commercial enterprise and not on a grant of public funds."

Gilbert put it differently: "There's been a completely erroneous attitude around that we're seeing basic research that has become of sudden, extreme value. Not true. Interferon is a perfect example. That basic research into interferon probably costs some fraction of a million dollars all together. The cost of the applied research in the *companies* trying to make it will be on the order of $20 million to $50 million. What happened in the last decade was rather the realization that this research *would* be of great commercial value" many millions of investment dollars down the road, "not that it *was*."

On paper, the commercial value is working out in stunning figures. After Genentech's landmark public stock offering in October 1980, Boyer was worth about $40 million. The stock price and thus Boyer's net worth have dropped some since then, but not enough to make him less than fabulously rich.

The effect on academe: electric. Good scientists have been willing to join up with the small firms because they dislike the corporate anonymity of the majors, about which one scientist said, "You're just a number in the parking lot and that's how your research rates, too." But certainly the financial rewards have not gone unnoticed. Says one molecular biologist now with a small rDNA firm: "Look at me, I'm 40 years old, I've spent my whole life in academia and I'm at Harvard, which is the top. And what do I get for it? To live like a poor student."

A small science community suddenly finds itself thrust into the limelight—and into a position to change the world. The combination is thrilling and unsettling.

Crossing a grassy square in a quadrangle of old brick buildings, you suddenly look at the life-size bronze rhinoceroses that mark the entrance to Harvard Biological Laboratories. Inside, the old labs offer no testimonial to the importance of their scientists or the power of their discoveries. Walter Gilbert's lab is here, at this time in late 1981. Down the hall is Mark Ptashne's. The two are said not to be on good terms. Up and down the halls, the talk is of companies. His company and her company. He consults for this one, she is starting that.

In an office near Gilbert's, Vicki Sato laughs at the business boom. "Everybody's got a company, I mean everybody." Sato's is Angenics, a firm she helped incorporate to put monoclonal antibody technology into the marketplace, and for which she will soon leave her associate professorship. "When we get together for coffee, the conversation is always what firm somebody has started or joined."

And there are wry twists to the new combinations. Sato was then dating Lewis Cantley, also an associate professor of biology —but he was with a different firm. "We've joked about whether conversation at the dinner table would suddenly stop as we realized we couldn't talk about what one of us was doing," she said. "But seriously, I don't think it's going to be a problem. It isn't around here, it's just a new fact of life." Sato and Cantley are now married and, yes, conversation was sometimes strained, until he ended his formal consultancy with the "other company."

Sato is a friend of Gilbert, and says he was a big help to her when Angenics was starting up. "He had already been through it all with Biogen and he offered a lot of good advice."

What about restrictions on the flow of scientific information? "I found it to be the opposite," she said. "Wally [Gilbert] and I talk about more science now than we used to. We talk about the work we're doing at our companies as well as the work in our labs. Sure, maybe at some point in the conversation we say something general or vague in order not to give away too much—but that's no different than when two scientists are competing in the same academic field."

No one is expected to give away trade secrets, and in any case, molecular biologists have never been inclined to give away the farm. At this stage in genetic engineering's development, the style of competition practiced in academia—the race to be out first, the race to use the good work of competitors to surpass them—may be among the healthiest signs for the growing industry.

Sato's business talk is a hardheaded variety that would warm the worried heart of a Wall Streeter. "We're trying to build a business," Sato said, leaning forward intently. "We're not trying to make a haven for basic science in the absence of good business."

Of her firm she says, "Hybridomas [the basis of monoclonal antibody development] are a mature technology. There's room for improvement and modification, but it's usable now at the industrial level, so it's easier for us to envisage products immediately producible without basic research" than for those concentrating in recombinant DNA to do so.

The proximity of men and women who are at the top of their field also leads to conversations sprinkled with asides on who's married to whom. Tom Roberts and wife, Gail Lauer, are both leading scientists in maximizing gene expression; Lauer specializes in the control regions of genes. Richard Losick, Harvard professor of biology, is married to Janice Pero, a former associate professor, and both specialize in the genetic mechanisms of *Bacillus subtilis*, a bacterial "factory" that operates differently from *E. coli*, with potential industrial advantages.

Maybe business is the new shop talk in the lab, Losick and Pero say, but it took time for them to become accustomed to the sudden "glamor" foisted on their pet bacterium, just a couple of years ago known in America only to industrial fermenters and academic biologists. Pero says, "We've spent our careers studying the control regions of *B. subtilis*, so it's always been important to *us*. But neither of us foresaw the possibility that what we were doing would have commercial value for decades."

Losick adds, "We were doing pure research, because it interested us. All of sudden my mother is asking me what stock she should buy. I think it's crazy, the way the industry has taken off."

Losick, like some other top scientists and executives, sees a bright future for the science and the technology it is spawning, but a future that will be enjoyed by only a few survivors of the current company-forming craze.

Nevertheless, Pero and Losick are both part of that boom. Pero recently left Harvard to join a Cambridge rDNA firm full time. Losick, who has tenure, will remain at Harvard, but will be a part-time consultant with the firm. And their feelings about such commercial work, long thought of as anathema to serious academics, echo those of Sato: They look forward to seeing their research benefit the public.

The bacterium Losick and Pero have studied is a soil-dweller rather than an animal parasite. That may mean that cloning genes into *B. subtilis* would avoid the hazards—real or imagined—of cloning deadly traits into a microbe that could carry them into humans or higher animals.

But in the race for production quantities it is *B. subtilis*'s structure as a factory for the new engineering that makes it such hot property. When *E. coli* makes a protein, the protein remains within the inner-cellular "soup." *E. coli* reproduces rapidly, a valuable trait in making large quantities of protein, but researchers must then add a chemical to burst the cell walls of the colony in order to get the protein, be it insulin or one of the alcohol-producing enzymes.

B. subtilis obligingly squirts many of its protein products straight through its cell wall into the surrounding medium, requiring only that the product be collected, a feature the organism may have developed because proteins secreted into the soil remain near the producing bacterium, so no inner storage area is needed.

A further advantage of *B. subtilis* is that the chemical industry has already done a great deal of fermentation with it. *E. coli* is a favorite in the laboratory, but far less known in industry—though both are outdistanced by yeast, that most ancient of fermenters, in its familiarity to industry.

For every microbe touted there are those expert in its manipulation. The uncertainty about which organisms will prove commercially most suitable means an equal uncertainty about whose work will pan out the richest. Losick points out that added to those unknowns is the even more uncertain, personality-dependent question of *who* will turn out to be the best cloner.

"These [rDNA] companies are going to depend on young Ph.D.s, and they're unproven. The proven people, the top scientists, aren't going to be bench chemists. They'll be guiding development, but they aren't the ones who'll be responsible for the production of the huge quantities of something, or for the first successful clone or something new."

It might be argued that that's what makes Pennant races, too, and there's always a rookie of the year. However, all those uncertainties explain a fact of investment life in genetic engineering: the major companies are putting down a variety of bets on a variety of types of enterprises, opening the opportunities to many different rDNA firms instead of just a few.

In any case, Pero points out that because the small firms are carrying the technology, she is able to go into industry and remain in Cambridge as a senior scientist holding equity, rather than having to join a huge drug firm that allows little scientific freedom and requires relocation away from academic centers. That is a major draw for her.

But The First Boston Corporation stock analyst Joyce Albers, one of the few analysts on Wall Street with a background in science (microbiology), counters that much of this newfound financial freedom for scientists is illusory. "Nobody has made any money cloning things. There's a lot of talk about who can do what, but nobody's done it. So what is the equity worth? It's all just paper right now."

Lynn Klotz replies, "Right. Not unless you chain yourself to your desk to make it work. You've got to turn out product to make the paper worth something, and that's why *who* is doing science in a company is so important."

Tom Roberts noted that "people-dependence" makes it especially important for the smaller companies to hire as many of the best people as possible, right away. "A small company has to do a lot of exciting science—its most exciting—in its first few years, to attract attention and prove itself."

Offerings of equity and large salaries are making that possible, so there should prove a connection between company valuation and real dollars. Roberts pointed out that of job offers he has had from rDNA firms, "The largest was three to four times the salary I make at Harvard." Luckily, he said, his wife decided to become a full-time scientist for an rDNA firm, and he will consult. "Otherwise I'm not sure I could've withstood the pressure."

Roberts adds matter-of-factly: "It took me a long time to believe that money would be made from products using these techniques. But I do believe it. There will be products, it is going to happen." And overproduction techniques, which will be the key for the time being, are now only a matter of technical refinement, he says. That is an indication of just how the race is heating up. The chase through the labs of universities and companies alike, for the better promoter, the better carrier of genes into bacteria or yeast—all the efforts "are remarkably similar," he said.

Lynn Klotz, thinking of the future for BioTechnica, draws the conclusion: soon—not next year, but soon—*lots* of people are going to be able to get maximum yields out of bacteria. For one thing, there is a natural barrier in getting a desired protein from an organism, ranging from 25 to 50 percent. That is, out of an *E. coli*'s total protein product, the most you can get of the protein you want is 25 percent or so. If the bacterium commits more of its energies to expressing a foreign gene, there won't be enough left over to sustain its own life. You kill off your colony. Some day, in other words, every drug company or chemical firm will be able to overproduce to that natural ceiling. Ominous for the champion cloners? Not necessarily.

It is worth recalling Eggers' warning: The new companies are competing less with one another than with the old technology. Whatever any of them do to put the new technology in place helps them all. They must make certain that the major companies need their techniques long enough for them to earn sizable revenues, and then be sure they are farther down the road when the big corporations catch up to that point.

The small firms are also racing the clock to prove themselves while the honeymoon with investors goes on. Sato, speaking of Gilbert's help to her, noted that "It's important that we help one another out, especially in the beginning, because we have to develop products that will compete in the marketplace."

Klotz pointed out that the intimacy of that small group of scientists could also ensure their survival as a whole—that is, the survival of the industry. "Should there be a severe shakeout of companies, we may help each other get through it."

For now, though, the lab-mates, collaborators and competitors are winding up to face one another in a whole new ball game, and many are the possible slips and complications in that changeover. In 1980, the U.S. Supreme Court decided that "novel lifeforms" could be patented, providing the underpinnings for the industry, to paraphrase a Genentech announcement. It also put new tangles in the webs of patent regulations between universities and scientists that we will try to unravel later.

The overproduction patent held by Harvard, for example, originally listed Roberts, Ptashne, and others, but omitted Keith Backman. The patent was later altered to include Backman. The patent, as is customary for work done at the university, is held by Harvard in the scientists' names, and Harvard will determine who gets a license to use it. Stephen Atkinson, executive secretary for the university patents committee, said Harvard generally grants license first to whichever company it believes can make best use of a technology. However, the university usually does not give out exclusive licenses and in the past has earned relatively little income from patent fees, always set at a modest rate.

But in a case like the overproduction patent, suppose there were applications for exclusive rights—Ptashne is in a different company from Roberts and Backman. Collaborators in science, one of them might be denied use of the process all of them worked on.

Atkinson agrees that such twists make for headaches the university must work out. But he pointed out, "If these three were in industry when the patent was granted, they wouldn't have been able to take the technology with them when they left the company, none of them, without the permission of the company. In this case it's possible we'd license all of them to use it."

Further, the musical chairs set up by venture capitalists and corporations bring understandable rushing, fumbling, reshuffling —though so far, chairs enough for everybody. Klotz initially had planned to join the Harvard-backed company, then talked with Ptashne when it was spun off as a wholly non-university concern. All he knew of the business, Klotz recalls, was what he'd heard of Ptashne's efforts at Genetics Institute.

Roberts also was considering joining Ptashne. Then Klotz suggested to Roberts that they strike out on their own. Physicist Tom Coor, who'd followed the high-technology path years before by founding an applied physics company in Princeton, immediately took Klotz to visit a venture capitalist—Morton Collins, president of National Venture Capital Association as well as engineer, mathematician, and consummate hardhead.

Collins upbraided Klotz and the fledgling "genies" for venturing into the business world and expecting to be handed blank checks.

"Gimme a break," Klotz recalls saying, after nearly twenty years in science, "I've only been at this *business* sixteen hours."

Told he would need to develop a business plan, Klotz thought, "I don't remember Mark Ptashne having a business plan. I thought he just spread a bunch of CVs [academic resumes] on the table, and his backers handed him $6 million."

"That was the level of our naiveté about capital and business," Klotz said. "That all seems like a long, long time ago, in another world." What followed in business was what had preceded in science: work, work, work. Many scientists followed through on the particular demands of business, reeducating themselves, committing intellectual resources as did Gilbert and Baltimore, some committing their careers to a technology steaming along for now on faith.

But there is a unity to the science-business fragments. If scientists appear driven in their research, their own words imply that the motivation is more a pull than a push—a sense of quest. Practical things are not necessarily part of scientists' quests, but there seems no contradiction in finding them in a scientist's repertoire. Practicality, on the other hand, falls into the "logically necessary" category on Wall Street, but that thirst for quest is often found there, too, where the favorite old slogan preaches: "No guts, no glory."

John Hunt, among the most unusual of the genies, himself offers a telling comment on the willingness of many people in high placcs to commit themselves. "We're right on the frontier," he said in his office at Princeton's renowned Institute for Advanced Study, where he was associate director. "The world after this is going to be a different world." Though not a scientist, he has spent his life among scientists. Author of two novels, he was nominated for a National Book Award for his *Generations of Men*.[6] A "kid out of small-town Oklahoma," he rose to become executive vice president of the Salk Institute in California, vice president of the Aspen Institute in New York, and vice president of the University of Pennsylvania. He spent twelve years in Paris and was director of operations for the oldest French foundation, helping to establish and run the Center for the Science of Man there.

He then went to the Institute for Advanced Study, leaving in early 1982 to become chairman and chief executive officer of BioTechnica International, another gene-splicer.

"The changes that will be wrought by the people who have made this leap, who are making it now, will not only transform society," Hunt said. "They will test society at its very center, at the definitions of what life is and what makes it valuable, of what is human, and what is permissible. That's why this is so important."

Genetic Engineering: The Science II

THE COMMERCIAL GENETIC ENGINEER AT WORK

In order to understand the wealth of procedures and tools genetic engineers have at their disposal, we should look first at how a natural organism controls the amounts of proteins it makes at various times in its life. *E. coli*'s daily diet may vary greatly, as does our own, because this bacterium can live everywhere from streams to animal intestines. In one place *E. coli* cells may be surrounded by glucose, in another a quite different food substance. Such different foods call for different enzymes to digest them into needed cellular components; on the other hand, if the cells are temporarily in a "starvation" environment, they have no need to produce any digestive enzymes.

To conserve energy, cells try to limit the making of enzymes and other proteins to times when they are needed. Further, some proteins will be needed in far larger quantities than others. Proteins involved in the construction of large cellular components, such as ribosomes, would be required in huge amounts compared to an enzyme needed only for an occasional chemical reaction in the cell. Such observations lead to some of the central questions in genetic engineering: How does the cell control how much of each protein it makes? And if we can find out, can we trick the cell into using these controls into making large quantities of *our* protein?

Molecular biologists already know a great deal about the control regions of *E. coli*, and they know that a lot of the control—but not all—takes place at the DNA level; that means it can be interfered with easily by genetic engineers. For now, we focus exclusively on control of protein production at that DNA level. Consider the fact that the more messenger RNA copies of a given gene-region are made, the more copies will be bound to the ribosomes, making that much more protein. Then how does the cell control the number of mRNAs made? By instructions coded in the *control region*, a portion of the DNA that sits just in front of the gene region. The control region contains information in that same four-letter (i.e., four-base) alphabet, which determines how often mRNA copies of the single gene are made. Figure 5 is a highly schematized sequence of steps from DNA to protein. We've pictured the DNA as a single horizontal line and have not articulated the sequence of bases, even though that determines the instructions.

Through the schematic microscope of Figure 5 we can see a gene (G), control region (C), and a ribosome binding site (RB), whose functions will be clearer shortly. Bound to the control region of the DNA is a molecule of RNA polymerase (the black rectangle in Figure 5a), the enzyme that makes the mRNA copy. How many copies of the mRNA will be made depends on how strongly it binds to the control region; stronger binding means more copies, because more RNA polymerase molecules will bind and go into action.

After binding to the control region, the RNA polymerase travels along the DNA molecule toward the gene region, as the horizontal arrow shows, making an mRNA copy of the DNA as it moves along. The copying ends when the RNA polymerase encounters a "stop" signal on the DNA sequence after copying the gene (Fig. 5b). Notice that the copying begins before the polymerase actually reaches the gene. That's because the first thing that must be transcribed is the site on the new mRNA that will bind to the ribosome, so the ribosomal machinery can read its information and make the protein molecules called for.

The next step (Fig. 5c) is critical. The mRNA binds to the ribosome at a particular spot *complementary* to the mRNA's own ribosome binding site. Without this very specific binding, no protein will be made. Now bound, the mRNA moves along the ribosome (Fig. 5d). As it does so, the sequence along the mRNA is

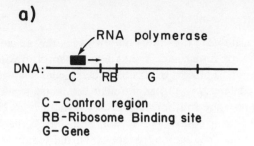

a)

RNA polymerase

DNA:

C RB G

C – Control region
RB – Ribosome Binding site
G – Gene

b)

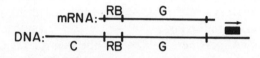

mRNA: RB G

DNA: C RB G

c)

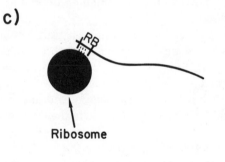

Ribosome

d)

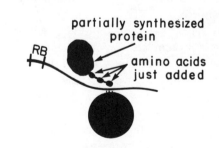

partially synthesized
protein

RB

amino acids
just added

Figure 5. The major events in the transcription and translation of a gene.

read and "translated"; the ribosome catalyzes the assembly of the right sequence of amino acids into a polypeptide protein chain, aided by certain adaptor or translator molecules called tRNA, to be discussed later. As the amino-acid beads are strung together at the ribosome, they fold into a globular structure, as beads on a string do if piled or rolled into a ball, and that is the completed protein molecule. You can also see a partially finished protein molecule, folded into globular form at the left as amino acids are being attached to the end of the chain on the right. We'll discuss protein synthesis in greater detail in Chapter 9.

Before rolling up the magnification on the schematic microscope, we need to look at protein manufacture in more detail and review what we've discussed so far.

1. The RNA polymerase binds to the control region in front of the gene. Whether the polymerase binds at all to this region and how strongly it binds depends on the base sequence of the DNA; in turn, the strength of the binding determines the rate of mRNA production.

2. The RNA polymerase encounters a start signal and begins copying the DNA as a messenger RNA until it reaches a stop signal telling it to quit. Both start and stop signals are encoded in the DNA base sequence.

3. The mRNA binds to the ribosomes, and another start signal —this one encoded in the mRNA—tells the ribosomes where to begin reading to make the protein encoded by the gene. At the end of the gene, after all the protein's amino acids have been assembled and linked according to the instructions on the mRNA, another signal is hit saying, "Terminate the amino acid chain": protein complete.

Now our earlier assertion should be clearer: All the instructions for making a protein are somehow encoded in the base sequence of the DNA and consequently in its mRNA copy. If we could just increase the magnification of our microscope so we could see how the bases were ordered in a particular DNA gene region—whether ATG or GGC, for example—we ought to be able to read in the language of genes. With enough samples of reading in this language, maybe we could come to recognize the control sequences for binding RNA polymerase, start and stop signals, and so on. In a nutshell, that is how the history of this area of molecular biology proceeded. The techniques used to determine the base sequence of

DNA were chemical rather than microscopic, but they accomplished the same goals. They allowed molecular biologists to read DNA sequences that spell command sequences for protein production or that correspond at the end of the production line to known proteins. This is what we mean by "sequencing DNA": learning the order of bases, which allows us to relate that order to commands or to protein segments.

The section of DNA control region where RNA polymerase binds and begins moving "downstream" toward the gene is known as the *promoter*, one of the control region's major components. The genetic engineer's bag of tools includes "strong promoters," so-called because polymerase binds strongly to them, and this efficiently makes an mRNA copy. In weakly bonded promoters, polymerase could diffuse away from the DNA before beginning a copy. We are beginning to understand what it takes to have a strong ribosome binding site on the mRNA, too. Using techniques for splicing DNA pieces, genetic engineers can build DNA pieces containing strong promoters and strong ribosome binding sites spliced in front of the gene for the desired protein. Finally, they can insert this "recombined" DNA into a host organism, which will make the protein in large quantities. This is one of the major tasks of a genetic engineer.

There is also another kind of "control" we need to explain. We mentioned that *E. coli* may find itself at some time needing none of a particular protein, then later needing a lot of it. Strong and weak promoters do not provide this kind of control. *E. coli* can regulate the amounts of *different* kinds of proteins made by having a strong promoter before the gene of one, a weak promoter ahead of the gene for another. But clearly this method does not allow for variations in the quantity of a *particular* protein at different times. These latter controls are known to exist and can be thought of as molecular switches, turning on mRNA production at one time, cutting it off at another. The switches physically block RNA polymerase from binding to the control region and from moving downstream from the promoter to the gene "start signal" by binding to sequences on or near the downstream end of the promoter region. These blocks are known as *repressors*, and those sequences between the beginning of the promoter and start signal where they latch on are known as *operator* sequences. *How* they operate will be part of our deeper look in Chapter 9.

THE GENETIC ENGINEER'S TASKS

Let's look at the six phases of the genetic engineer's work in some detail; along the way a lot of the phrases used in newspaper and magazine accounts of discoveries will become clearer. (1) Obtain the gene or genes for a desired protein or proteins; the usual methods are called "shotgunning," "copy DNA cloning" or "DNA chemical synthesis." Which procedure is used depends on the organism from which the gene will be taken and the number of base pairs that make up the gene. (2) Then, build a plasmid or virus DNA to be the gene's carrier or "vector"; it must contain the right control sequences (promoter, operator, ribosome binding site) for the host organism that will be making the protein. (3) Insert the gene obtained in (1) downstream from the control sequences in the vector. How far the gene is from the control sequence is also important. With the right control sequence and distance, the protein will be made in large amounts in the host. (4) Now, insert the rebuilt vector into the host, which will treat the new DNA as its own. (5) Clone a colony from the redesigned bacterium, a fairly simple and time-tested practice. (6) Finally, carry out further genetic engineering to counter unexpected problems, to increase protein production, and to help fermentation engineers scale up the growth of the organism to profitable commercial levels.

Take a look at the first step. Obtaining a protein's gene is usually easy in the case of bacteria or viruses, but often is very difficult with higher animals and plants. Shotgunning is the most frequent method of getting the gene. Here, *all* the DNA from an organism containing the sought-after gene is cut up along with thousands of others. The DNA is snipped into thousands of pieces just larger than gene size, using special enzymes called *restriction enzymes* that cut DNA only at specific sequence sites.

Next, *all* of the thousands of pieces are spliced into plasmid or virus DNA vectors; the vectors must be ones which can be carried in a growing *E. coli* culture. The splicing is done with an enzyme called *ligase*. The opposite of a restriction enzyme, ligase can bond together pieces of DNA which have been cut; it is the "sticky-end" enzyme of Chapter 3. These procedures are directly analogous to a film editor's cutting and splicing. The discovery of the restriction and ligase enzymes made recombinant DNA technology possible. If the splicing is done right, at least small amounts of protein will be made.

Look at shotgunning through the schematic microscope (Fig. 6, page 74). At the upper left are four identical *E. coli* cells, showing only the plasmid DNA in each, labeled P. The single hash mark on each plasmid represents the specific sequence recognized and cleaved by the restriction enzyme chosen for this shotgun. At the upper right is a single cell from the organism that naturally manufactures the protein we're after, showing only the cell's chromosomal DNA. The hash marks on the chromosomal DNA represent the four cleavage sites recognized by this same hypothetical restriction enzyme. After snipping open the DNA with this enzyme, we'll have pieces A, B, C, and D, and it happens that piece A contains the gene we're after, illustrated as a thicker line of DNA.

Our aim: to splice the piece of chromosomal-DNA containing gene A into the plasmid DNA of one of the four *E. coli* cells, then reinsert the DNA plasmid with gene A *back* into the *E. coli*, and grow up a culture of that *E. coli* to get large quantities of the gene.

The first step is to get both the plasmid DNA from the *E. coli* cells and the chromosomal DNA from the donor organism: there are routine chemical procedures to break both free. Next, all the DNA is snipped with the restriction enzyme, resulting in plasmid DNA with single breaks in them and four pieces of chromosomal DNA from the donor—one containing the desired gene. This whole procedure takes only a few days.

Next, we combine the restricted DNA from the plasmid and that from the organism, splicing them together with the ligase enzyme. This nets four plasmids, each containing a different piece of foreign DNA. All the plasmids now are reinserted into *E. coli*.

You might wonder why the donor gene is not inserted directly into *E. coli*'s chromosomal DNA—simply "stitched" into a break at the proper place in the chromosomal sequence. The main reason is that it is much more difficult to do this stitching inside a cell, compared to outside the cell on a plasmid, and plasmids already have ways of working their way into cells, where they generally express themselves independently of the cell's chromosomal DNA. The mechanism exists because plasmids are naturally exchanged among bacterial cells. Also, plasmids are generally *small* pieces of

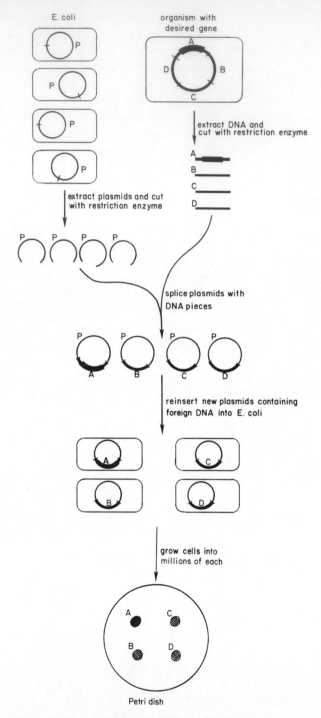

Figure 6. The procedure for "shotgun" cloning of a gene.

DNA, on the order of several thousand bases, whereas chromosomal DNA is in the millions of bases. Small pieces are easier to work with and analyze for results, making plasmids a good vector for putting foreign DNA into *E. coli.*

However, the DNA of certain bacterial viruses is also useful; these are viruses which attack bacteria, just as some viruses attack humans and other plants. Remember that in its normal mode, a virus attacks by inserting its DNA into a bacterium. Sometimes the viral DNA divides along with the infected cell. In this sense, it resembles a plasmid, although it doesn't usually exist as a little circle of DNA inside cells, but is often spliced right into the main, chromosomal DNA. Viral DNA falls into the same size range as many plasmids, so genetic engineers will sometimes use it as a vector. Looking back at Figure 6, note that the vector must be reinserted into the *E. coli* cells, whether it is plasmid or virus. In the lower center, four *E. coli* cells each contain a plasmid bearing a different piece of the spliced foreign DNA.

Two steps remain to get large quantities of gene A: separate the four *E. coli* cells and grow up millions of copies of *each*, then pick out the copies that have gene A. Separating and growing copies of each *E. coli* cell requires only that classical biological procedure, cloning. Ordinarily, although each bacterium in a group would give rise to its own daughter generations, the bacteria would grow all mixed together; we want them to grow up separately, as cloned colonies. In cloning, we dilute the *E. coli* in solvent to such an extent that there is about a centimeter between any two cells, then we lay the cells down on a gelatinlike substance called *agar.* As with gelatin, agar is solid but contains large amounts of immobilized water; thus the cells will have water but won't be able to swim and mix, and will grow on nutrients in the agar. In time, what was one cell will become a visible, usually white spot of several million cells on the surface of the agar. The bottom of Figure 6 shows a petri dish containing four spotlike clones; one of them contains only cells bearing gene A, one contains only B, etc. Which is which? We could extract DNA from samples in all four colonies and sequence it until we found gene A, but that would be a Herculean task. Our schematic microscope shows only four pieces of DNA in fragments; a typical experiment would have a thousand or more. Even worse, from that huge collection of DNA we would be hunting for one tiny gene region, a needle in a haystack. Sequencing just one piece might take weeks, so this project could last decades.

But after all, the gene we're looking for expresses itself as a protein. Why not find the gene by locating the protein in the only colony that will be making it? Often simple chemical tests will color the colony containing the right protein. In our experiment, the clone containing the desired gene is black instead of cross-hatched. For example, there is such a color assay for the protein β-galacto-sidase. If *E. coli* cells contain it, the assay will turn the entire colony blue instead of white—a difference that makes finding the right clone among thousands a snap.

Earlier, we mentioned that inserting a foreign gene into a bacterium is easy compared with getting the gene to express its protein, and clearly that is what we need to do to use these assays. Getting *E. coli* to produce the protein encoded by a gene requires the genetic engineer to do some advance work of the kind we've just described. He or she builds a plasmid containing a single restriction site at a point slightly *downstream* from a promoter and ribosome binding site, because they must be read out by the RNA polymerase first. The plasmids shown in Figure 6 are not natural ones; they represent artificial plasmids built up over the past few years by genetic engineers, each with different promoters, ribosome binding sites, and sites for many different restriction enzymes. Each of many different restriction enzymes recognizes a different four- to eight-base sequence. The genetic engineer pulls out one of these special plasmids, depending on the shotgun experiment planned, then goes to work to break out chromosomal gene fragments. This is just one of the procedures used to clone and identify a particular gene.

"Shotgun" was coined to describe the random nature of blowing an assortment of DNA bits into a group of plasmids and then shooting the plasmids back into *E. coli*. Doing all that and cloning colonies is not difficult; the steps are routine and the procedure takes from a few weeks to a few months. Most often the hard part is finding a simple assay to *locate* the clone containing the desired gene.

In the absence of a color assay, one might try to identify the protein by its function. For example, since interferon protects against animal viruses, animal cells might be injected with serums of each colony to locate the one which provides immunity against virus. You can do such a test with thousands of small colonies, but it is tedious work.

Another method of identifying proteins is to use flourescent or radioactive antibodies. The right antibodies would bind only to the sought-after protein, and their label could be easily read by laboratory instruments. But obtaining specific antibodies can present major difficulties in itself, so this method may not be usable. In some other cases, large quantities of messenger RNA for the gene being hunted can be obtained. Similarly labeled, the mRNA will bind to the DNA on the right gene region and can be detected with instruments.

In some cases, identifying proteins after shotgunning could take years instead of months, and that is certainly something for a research director of a company to consider before contracting with a genetic engineering firm to do a protein project.

Often the major advantage in shotgunning is that by the time the proper clone *is* detected, the colony is already making the protein; the genetic engineer may only have to fiddle around with the exact placement of the gene with respect to the promoter in order to achieve overproduction of the protein. Once the gene is identified, it can be snipped out, placed in any properly "tuned up" vector, and inserted in a final host for commercial production. At this point, we've covered all but the last part of the genetic engineer's work, but that may be the most difficult of all—preparing for commercial production.

This last phase might require radically increasing the amount of protein produced, controlling that production through the use of the proper promoter and operator sequences, finding a better host organism and engineering *it* to do the job, and adapting the whole procedure to large-scale industrial fermentation. One wonders if it is sometimes not "deliberate oversight" when firms announce major breakthroughs, even though they've failed to achieve a high level of protein expression in an organism suitable for industrial or pharmaceutical fermentations.

We also noted that obtaining genes from animal or plant cells can be very difficult, taking many years in some cases. It's easiest to get genes for small proteins of, say, fifty amino acids or less, compared to large proteins that can possess upwards of several hundred amino acids. Hormones are a potentially interesting class of protein, and usually small. They dramatically affect many bodily processes: absence of human growth hormone, for example, causes dwarfism, and absence of insulin can kill.

Genes for small proteins such as hormones can be chemically synthesized. Instead of pulling the gene from the chromosomal DNA, it is synthesized from the four bases—A, T, G, C—which are commercially available in a chemically active state and are not expensive. This synthesis can be carried out manually or using one of the highly publicized gene-machines.

Making synthetic genes for 50-amino-acid proteins requires only that 150 bases be strung together in the right order. Such a synthesis might take a few months. Right now the practical limit for gene synthesis is about 200 bases and a few months' work, but that results only in proteins of under 70 amino acids. Future improvements may make gene synthesis for larger proteins much more practical. But the few months involved are brief compared to the time required in the alternative method of copy-DNA (cDNA) cloning.

The relatively easy and reliable shotgunning technique does not work in obtaining most genes of higher plants and animals, and at the moment so-called copy-DNA cloning, discussed in Part III of the science section (Chapter 9), is the only way to obtain genes for these larger proteins. The reason is the existence of a strange trait in these higher organisms: interspersed among the genetic information coding for protein are gene regions that can only be described as gibberish. Organisms that contain these "nonsense DNA" regions also have means of recognizing them, and they are thus eliminated from the mRNA before it reaches the ribosomes to begin protein manufacture. Bacterial genes don't contain any gibberish; therefore they have no way of eliminating it.

Clone an animal or plant cell gene into a bacterium, and you get a protein whose chain is interspersed with incorrect amino acids coded for by the gibberish regions. The protein would likely be nonfunctional. Thus, cDNA cloning requires an editing process; the end result is a piece of DNA with the gibberish excised. The procedure is very difficult in almost all respects, and we'll just highlight its major points for now. First, the genetic engineers must pick up on the mRNA after the point at which its gibberish has been removed. They then use an enzyme known as reverse transcriptase to make a DNA copy *from* the mRNA. Notice that this is exactly the reverse of RNA polymerase, which makes an RNA copy from a DNA region. As mentioned in Chapter 3, David Baltimore's Nobel Prize was for, among other pioneering work, the discovery of reverse transcriptase.

The complexities of cDNA cloning are many. It's often very difficult to obtain the necessary mRNA; making the cDNA copy is not easy; and finally, further preparation of the cDNA copy is needed before it can be inserted into a plasmid vector. The whole procedure could take years for a particularly hard-to-find or particularly large mRNA. In the final science section, we'll talk a little more about cDNA cloning and gibberish-containing genes.

At any rate, let's suppose that the genetic engineer—by hook or by crook—has the gene he or she wants cloned in *E. coli*, has identified the clone colony containing the gene, and has succeeded in getting some level of protein expression from the gene. If engineers are very lucky, protein is being pumped out at a high level. Most likely, they don't get that lucky and have a great deal of work cut out to achieve a good yield. Also, *E. coli* may not be the host organism the genetic engineer ultimately wants to use for commercial production. Once most engineering problems have been solved, they may want to develop *B. subtilis*, yeast, or other microorganisms for large-scale commercial fermentation.

To get a higher protein yield, the genetic engineer has several options. First, any way that increases mRNA production will increase protein. To accomplish this, the engineer might increase the number of copies of the plasmid containing the gene and control sequences. Plasmids can exist in cells in multiple copies. The number of copies depends on one of the plasmid's DNA sequences, a signal that tells the host cell to make another plasmid copy. This signal region is called the origin of DNA replication. Remember that strong promoters help begin transcription of mRNA. Analogously, if the origin of DNA replication is strong, the plasmid will reproduce more often, and more copies of it will be available. Recently, genetic engineers have isolated strong origins of replication from plasmids, identifiable because the plasmids containing them are always found in many copies in the host cells. These origins can be isolated and spliced into a quite different plasmid that contains the desired gene. More plasmid copies, more mRNA, more protein.

Another method of increasing protein yield is to find strong promoters, as mentioned earlier. A relatively simple way to do it: locate the proteins produced naturally in the largest quantities in *E. coli*, identify their coding genes, then search "upstream" from those genes for what ought to be strong promoters. By studying promoters, engineers can learn what makes a promoter strong or weak. Then they can splice parts of various promoters together and arrive at a hybrid promoter stronger than those found in nature. A few such artificial strong promoters already have been constructed in *E. coli* to increase mRNA synthesis for those messengers that lead to important proteins.

Still a third method of increasing protein production is to build better ribosome binding sites on the mRNA. The sequence of the basic ribosome binding site for *E. coli* is known already, but there is increasing evidence that the base sequences surrounding that site not only affect binding but may also affect the *rate* at which translation begins.

Combining all these tools and tricks of the genetic engineer's trade, production of a desired protein can now be increased to 20 percent of the total cellular protein output. Consider that *E. coli* probably contains over 1,000 different proteins, and that an average protein therefore amounts to 0.1 percent of the total. That means genetic engineers are able to get yields of their own proteins ranging up to two hundred times that of the normal cell.

There is a catch. Even when all the promoters and other aids are perfectly placed, some proteins are not expressed in quantity. Sometimes they are degraded (destroyed) as they are made by enzymes called proteases, which are present in all cells. For some reason, *E. coli*'s proteases degrade some genetically engineered protein products before they can be isolated from the cell, but not others. As a rule, it seems that smaller proteins such as hormones tend to be degraded faster. Consequently, experimenters have been trying to develop *E. coli* strains that are protease deficient, so that degradation of the product will at least be reduced. At present, 70 percent of the proteins engineered into *E. coli* can be expressed in large amounts; for a variety of reasons, 30 percent cannot be, which is important to keep in mind in any given genetic engineering project.

OTHER MICROFACTORIES

We conclude our discussion of the fundamental science of genetic engineering with a look at two other interesting organisms that may become rDNA factories, *B. subtilis* and yeast. They are interesting primarily because of valuable enzymes and pharmaceuticals they can help us produce.

Consider the previously mentioned industrial enzyme α-amylase, which turns cornstarch into sugar that can be fermented into alcohol. Genefacturing the enzyme in *E. coli* requires destroying the colony to get the product. But in *B. subtilis* the α-amylase will be secreted out of the cell into the surrounding medium. We would then spin the cells down in a centrifuge, concentrating the cell solids into the bottom of the centrifuge tube, and pour off the liquid with the nearly pure enzyme, saving a lot on purification costs.

The general principles of genetic engineering remain the same unfortunate because it means all the strong promoters collected in in these other organisms, although details differ. For example, the promoter DNA sequence from *B. subtilis* will work in *E. coli*, but those from *E. coli* will not work in *B. subtilis*. That is especially recent years for *E. coli* cannot be used in *B. subtilis*. Nevertheless, scientists have begun to isolate strong promoters from *B. subtilis* and to learn why promoters differ between the two organisms. Ribosome binding sites also differ. However, while it took many years to develop sophisticated genetic engineering technology in *E. coli*, it should only take a few years to develop parallel technology for *B. subtilis*.

Even more importantly, we will have to learn from scratch how *B. subtilis* secretes its proteins through the cell membrane into the medium. Apparently, *B. subtilis* secretes certain proteins because they carry amino acid "signal sequences" that are compatible with the molecules of its membrane, allowing the protein to enter the membrane and pass through it to the outside medium. Just before exiting, these signal sequences are usually sliced off, so scientists do not know what many of them are. Once known, the signals could be tacked onto genes for almost any protein so that they too could be secreted into the medium uncontaminated by the cell's other proteins. This could make *B. subtilis* the host organism of choice in industry.

Yeast, of course, is an alcohol producer. In addition to the value of alcohol already discussed, genetically engineered yeasts may be used in the alcoholic beverage industry to make so-called "light beer." Regular beer contains so many calories partly because the starch from the grains fermented into beer cannot be metabolized by yeast, which lacks the proper enzyme. Unfortunately, we don't lack the enzyme, so when we drink beer, we metabolize the starch into high-calorie sugars. Result: the infamous beer-belly.

It should be relatively easy to genetically engineer starch-degrading enzymes into the yeast along with promoters that will assure quantity production of them. Difficulties follow: brewer's yeast may not secrete some enzymes into the medium to degrade the high-calorie starches, so even if the enzyme were produced it might not get to where it could do its job. But the problem is not insoluble. Some yeasts *either* partially ingest starches, so secretion of the enzyme would not be necessary for the enzyme to work, *or* some enzyme secretion does take place. Further, it also should be possible to overproduce the enzymes in bacteria separately from the brew, then dump them in. Then the yeast will eat the resulting sugars, so it, rather than you, will get fat.

So far, we've only talked about yeast's production of alcohol, but it could also be used eventually to make products that cannot be made in *E. coli* or *B. subtilis*. Certain proteins of higher animals *and* yeast undergo modifications after translation on the ribosome. For example, some proteins have sugar residues tacked onto them in various spots. Bacteria never make these so-called "post-transcriptional" modifications and don't have the apparatus to do so. That means any animal protein that requires these alterations cannot be made in bacteria. Some interferons undergo such changes, but it isn't clear if the changes are vital to the interferons' function. Nevertheless, yeasts offer the possibility of making these proteins with their modifications. We still aren't sure they will be able to make the changes in the right places, but we should know within two years. Yeast genetic engineers, too, have located several yeast promoters and are beginning to understand their operation. Other yeast signal sequences are also under study. Thus, the DNA language of yeast should be easily "readable" within two years.

If so many problems loom in trying to engineer animal genes into bacteria, why not forget bacteria and go right for the animal cells as host? Why not place interferon genes within animal cells in front of strong promoters to make interferon plentiful in its native habitat? Major and very recent advances may make this possible. Genetic engineers can now greatly engineer animal cells to make even foreign proteins. But animal cells are very hard to grow outside the animal in the laboratory, and they grow very slowly.

Cells of higher animals grow in a very protected environment, bathed in nutrient-rich blood and maintained most of the time at well-regulated temperatures. It is possible to alter them so they will grow in a nutrient medium in a laboratory dish, but they are very finicky. Their growth medium must be rich in vitamins and nutrients, and that means it is costly. *E. coli* will grow well on sugar water with a few additives. By contrast, the standard medium for growing animal cells is the serum of fetal calves.

Growth rate is even more serious a problem. *E. coli* and *B. subtilis* double every twenty minutes, but an animal cell doubles once every twenty-four hours. That means in ten to twenty hours you can grow up a fermenter-full of billions of *E. coli* cells that might yield hundreds of pounds of protein. The animal cells will have doubled just once in that time, and it could take weeks to get a significant yield of protein product.

But the real potential for engineering directly in animal cells is therapeutic. Lack of certain proteins or mutations of existing proteins causes many genetic diseases, including sickle-cell anemia, Tay-Sachs disease, and β-thalassemia. If you could engineer into the patient the gene coding for the protein, you might cure the disease. Certainly, there are scary aspects to this kind of genetic engineering: you might alter humans for other than therapeutic purposes. We have some time before we need to worry about this—although recent experiments in which rat genes "expressed" themselves in mice certainly brought us closer to this type of "fancy" genetic engineering in living animals.

The advance of animal-cell genetic engineering will probably be much slower than that in yeast or bacteria, despite some discoveries concerning promoters and other sequences. Added to the growth problems, it is difficult to keep animal-cell cultures uncontaminated by bacteria or yeast, which overwhelm them in numbers. All these factors tend to slow research in this area.

Plant genetic engineering is likewise in its infancy. Little is known about plant promoters, and work is just beginning in the development of the vectors needed to place engineered genes into plant cells. Nevertheless, the social and economic potential for plant genetic engineering is staggering, if not at hand. As it develops, we are also going to have to consider the hazards suggested by some scientists, such as the possibility of altering nature's balance.

chapter 5

Bull, Bear, and Clone

VENTURERS

Life generally proceeds by small changes rather than by cataclysm, whether the changes are planned or accidental. An alteration in the way things usually occur will produce an effect, small or large, and life will either follow the new course as a superior/easier one or ignore what then becomes aberration. Form follows function, for the "best" form among a large number competing for acceptance is the one that *works* best.

That might pass for a business view of evolution or an evolutionary view of business: orderly, nonradical, a survival-of-the-fittest scheme in which corporate form follows product demand, in which laboratory discovery takes root in pilot project, then blossoms into a "big boomer" in the test markets to yield an industrial giant—or is ignored and never comes to be at all.

Now consider genetic engineering. The largest collaboration ever between business and biology, the creature was born in direct violation of virtually any evolutionary notion. The industry was born fully formed. More than a hundred companies, set afloat on a combined investment of more than $500 million, work at a fever pitch to fulfill a demand that is so far only envisioned; the first products of genetic engineering are just entering the market.

Further, these companies are ultimately depending not on orderly growth, but on a super "score" that can only come to some

—and without which the prospects for survival range from slim to none.

Is this any way to do business? Maybe so, because this industry was built by venture capital, and venture companies tend not to be born, live, or die the way other firms do. Significantly, much of the controversy in the investment community over genetic engineering revolves around just these questions of how an industry should properly come of age.

Venture or risk capital is a largely American institution, although the notion runs through most free enterprise economies. It refers to the money you can afford to lose, money you can therefore bet on something with long odds, but which would yield a really big score if the project connects.

Thomas Perkins takes some exception to that common definition, but in so doing he puts a finer point on it. The chairman of the board of Genentech and the first venture capitalist to put money in the field, Perkins says: "Venture capital does not go after high risks. Venture capital seeks high returns. High risks just tend to go along."

The sources of venture capital are varied, ranging from the wealth of America's most powerful families to some corporate funds that are plowed back into the field—instead of into taxes, for example.

Perkins himself is a perfect illustration of how things get started. A graduate in electronics from M.I.T. with a Harvard MBA, Perkins rose in the corporate ranks of the budding electronic data processing industry. He started Hewlett-Packard's computer division, then became its general manager. As part of his job, he oversaw the creation of Optics, Inc., a firm in the even-newer field of laser technology. There he got a bright idea and brought it back to H-P with him. He conceived a laser that would lack the fine-tuning refinements of laboratory models, but which might be used for "dirty" jobs—inspecting pipes on construction sites, for example—jobs that would involve wear and tear, but would not require the precision of the scientist's tool.

He remembers asking H-P "if I could carry on my little hobby on the side, of developing this idea I had, and they said I could. That was pretty unusual for a company to grant to someone at my management level."

Looking back, Perkins says he was at that moment a venture capitalist, but didn't know it. "What I did was to assemble the people and the money to bring this idea to reality," Perkins said. The result was what he calls the lasertron, the world's first "light-bulb" laser that simply switches on and off without tuning. "We did very nicely with that," he says modestly. He is a senior partner in the San Francisco venture capital firm of Kleiner, Perkins, Caufield and Byers, whose spacious offices overlook the Bay and Marin County from the upper reaches of an Embarcadero tower.

A soft-spoken man, he says with more than a touch of pride, "If we hadn't backed Genentech, I'm not sure the industry would have formed the way it did. I'm not sure where it would be now."

Genentech is not the oldest of the genefacture firms, but it was the first founded purely to do recombinant DNA manufacture: the central, most elegant, and in the long run, most promising art of biotechnology, and what we have been referring to as genetic engineering.

Venturing, basic version: man with capital meets scientist with idea. Realizing that each has something the other can use, they put the one's cash and the other's superbugs together and hope for the big one: the patentable bug that spits a miracle drug all day and multiplies like crazy—or some variation on this. Herbert Boyer met Robert Swanson of Perkins' firm that skyrockets went of genetic engineering, but there was still no industry. It was when Boyer met Robert Swanson of Perkins's firm that skyrockets went off. Swanson and Boyer each put up $100 of their own money to incorporate and Swanson hit Perkins for $100,000 in start-up money. "He was hot as blazes on genetic engineering," Perkins recalls. "He was going to make a career out of this either with us or without us."

Risk? "Very high. I figured better than 50–50 we'd lose it. But it's rare when the odds on a new technology are better than 50 percent."

Second thoughts? "Not at all. If it worked, the rewards would be obvious."

Boyer and Swanson had a strategy for quickly developing products such as insulin, human growth hormone, and other drugs that could be sold to existing major drug companies with ready markets, so it was no accident that the company teamed with Eli Lilly to develop human insulin. This short-term revenue-producing plan was to be matched with a longer-term strategy of becoming a fully integrated pharmaceutical company—researching, developing, marketing. Finally, the two saw Genentech farther down the road as a challenger in the capital-intensive field of manufacturing industrial chemicals not now made through biotechnology, a more revolutionary use of genetic engineering they felt would be unlikely to produce immediate revenues.

At the moment Perkins was giving Genentech its $100,000 launch, on the other side of the country Robert Johnston was getting excited about the field. A former investment banker who had been putting up money for and running various medical equipment firms in a business run from his Princeton home, Johnston was in regular contact with biochemists, molecular biologists, and medical researchers.

"I guess what I couldn't believe was how excited *they* were by the technology," Johnston said. "Scientists, you know, are a pretty conservative lot, especially when it comes to guessing how practical the applications of their discoveries will be. To some, it's almost a dirty word, 'practical.' But all of a sudden, everywhere I went, *scientists* were enthusing over genetic engineering and what it could do."

That period, the late 1970s, also marked the height of public controversy over genetic engineering and its feared risks. Many of the questions raised about recombinant DNA were valid. But, not surprisingly, a heat storm erupted around the legitimate questions simply because threats are perceived in *any* new technology. In that adversity, entrepreneur that he is, Johnston saw opportunity.

"The major pharmaceutical houses were holding off [putting the technology in place] because of the controversy," he recalled. "Pfizer, one of the country's biggest houses and one of the most experienced at fermentation technology, had been picketed in Connecticut, and they had stopped in their tracks. It seemed to me a good chance for smaller companies to step in, develop the technology, and sell some useful products competitively."

Unlike Genentech, Johnston planned to develop genetic engineering techniques for producing amino acids, for which there was already a large, lucrative market. These protein building blocks, as we pointed out, are useful nutritional supplements, especially in animal feeds. They were already being produced by microbial fermentation, so to enter the market one needed only to build a better microbe. Johnston had no microbe engineer, but what is an entrepreneur if not enterprising?

He put an ad in *Science* magazine, the official organ of the American Association for the Advancement of Science. "Entrepreneur seeks scientist to be president of genetic engineering company," the ad read, then went on to detail plans and requirements. Johnston got about half a dozen replies, but "most of the respondents were unqualified."

Only one applicant completely filled Johnston's bill of particulars—Dr. Leslie Glick, who not only became Genex's president but later founded the trade organization serving biotechnology companies. Johnston says, "You usually think of mad scientists and conservative businessmen. We're kind of a reversal—mad businessman and conservative scientist." It was Johnston who had to convince a doubting Glick of the enormous promise in the field.

The nearness of Genex, in Rockville, Maryland, to Washington aided the company in becoming a major collaborator in the federal Office of Technology Assessment study on the future of applied genetics. Stock analyst Scott King, who helped draft the report, says that the collaboration was too cozy, given the company's interests in the field, but a Genex board member counters that the firm was the only one that had the technical resources to provide the data the government needed.

Johnston recalls hard early days, a year in which the company existed virtually only on paper as he and Glick tried to scrape up funds. But he sees that as a boon to the company's corporate toughness. "It's better to grow up lean, appreciating the value of a dollar," Johnston says. "A lot of computer companies in the sixties went down because they got way too much money when it was plentiful, developed bad habits, and could not survive when the crunch hit."

That crunch, totally unknown to or forgotten by most people, is a major landmark on the rough economic terrain of high technology, and is frequently referred to by those now assessing genetic engineering. Like biotechnology, electronic data processing (EDP)

—the computer industry—was born on waves of futurist fanfares that brought millions of "easy" investment dollars before the bubble burst in the early 1970s. A few of the best EDP firms survived into this "golden age" and are now making a fortune, but hundreds of others did not. Many observers foresee a similar chasm yawning between biotech and its golden age.

Genentech has been a public firm for several years, but Genex has spent most of its corporate life privately financed, and there is a vast difference between public and private financing—each having its pros and cons. Johnston pointed out that public money "is the cheapest money available." The corporation pays far less back in earnings on the public investment dollar than it must give away in equity in return for venture capital or private financing arranged through investment bankers. Investment bankers broker stock offerings to private investors, but they generally do not take part in managing the company. Venture capitalists do.

But public money has drawbacks. Morton Collins, a venture capitalist who survived the computer crunch, notes, "Public money is impatient, demanding, and volatile. When the public likes you, the money flows. But when the public and the Wall Street analysts decide it's time you showed a profit, the money stops quicker than any other kind."

Whether public or private, most genetic engineering firms have venture capital somewhere in their financing history, and that in itself is new. Collins, president of the National Venture Capital Association, a 117-member trade organization and lobby, said that the concept only solidified in the last fifteen years, and that at its inception it was not at all specialized with respect to areas of investment. In those early days, Collins said, he foresaw a need for seed money in the booming data processing industry—and if there is a phrase that describes venture capital best, it is "private seed money." Collins founded Data Science Ventures at about the same time two friends were trying to start a similar "seed money" firm to back companies specializing in medical systems. The time was not yet ripe for Peter Farley and Ronald Cape, Collins relates, but they went on to develop Cetus.

But venture financing not only brings money, it brings management controls. Entrepreneurs such as Johnston typically put together business plans, then look to venture capital for financing.

A given firm might offer $2 million in return for half the company, *and* provide a certain number of directors for the firm, possibly the board chairman. By these actions, the venture group has put a value on the company based on its experience of corporate worth, and it provides expertise in running the business as well as the infusion of money that starts it off.

Collins says, "We've done more than fifty start-ups in fifteen years. When we come in, our executive expertise is worth more than the money we're bringing, and we try to impress that on those we deal with."

In their turn, what venture groups think is of highest value in the companies they back is personnel—particularly if the company represents high technology, the area in which venture groups have done most of their financing. Johnston says, "Most venture people will tell you management is 80 percent of the worth of your company. That means business executives and your top scientists."

Venture groups say they also bring a measure of stability to public offerings, in the interest of protecting themselves. Gerald A. Lodge, whose InnoVen had its own key role in the industry's creation, pointed out that the Securities and Exchange Commission has relied on adequate disclosure of corporate information to protect the stock-buying public. If venture groups were to sponsor offerings that turned up consistent losers, tougher regulation would be inevitable, "and that's the last thing we're interested in."

Lodge is chairman of InnoVen, perhaps the only venture group in which corporate investors play a working role. In 1974, as the industry's day was just beginning, Lodge, a partner in what was then a small venture management group, was contacted by the Monsanto Corporation and Emerson Electric. He was asked to "meld our capital management with their technical expertise" in a series of investments of corporate funds and some private investors' money. Lodge agreed, and a year later found himself at a Monsanto tutorial that focused on genetic engineering.

"The gist was that this was the field to watch, but no rush about it. The corporations didn't think we'd find anything to invest in for several years and believed the technology would not yield sales and profits for another ten years."

Lodge, then on the venture association board, ran into Perkins at a meeting only a short time later and heard about Genentech.

"I want in," Lodge said, and InnoVen made two large investments in the company as it was starting up.

Monsanto executives were delighted with the investments for two reasons—they saw a "window" on the future, and they foresaw a necessary hard learning experience on the horizon. "They said they were glad we got in so early, but basically: 'You're going to lose your ass. The start-up funds will be liquidated before the company makes a nickel.' "

That forecast, like so many others in this field, proved greatly mistaken. By 1981 Genentech had produced revenues of an estimated $11 million on contracts with Ely Lilly for insulin, Germany's KabiGen for human growth hormone, and Hoffman La Roche for interferon. The firm reported earnings per share—$.02 —for the first quarter of 1983, and its stock was selling at $42.25 in early May.

But Lodge remembers a "uniformly conservative view" by most scientists as recently as five years ago. A topflight medical researcher gave Monsanto a list of eighteen technical hurdles that had to be overcome before recombinant DNA would have an impact on industry. He projected a five- to fifteen-year time lag for every product, based on when the various hurdles would be cleared. "All eighteen challenges were met within two years," Lodge said, an indication that the products are much closer at hand than originally forecast.

In the summer of 1977, Johnston and Glick approached Inno-Ven for funding. The "storm and heat" over hazards were then at their peak and Lodge initially was hesitant, but the firm eventually put up "several hundred thousand dollars" for 60 percent of Genex. Lodge made it clear that he was interested in Genex's ability to apply genetic engineering to industrial rather than pharmaceutical products. In effect, this meant Genex was to prove its skills at reducing costs of making existing chemicals rather than concentrate on new inventions such as wonder drugs. This illustrates the management side of venture financing.

Despite these pluses of private financing with its more savvy investors, it is worth noting that all the major genetic engineering firms have now gone public. Genex offered two million shares in September 1982, and Biogen sought to raise $60 million for 13 percent of its stock—2.5 million shares—in a public offering in March 1983.

SHAKE-OUT

Mort Collins lives just down a country road from Bob Johnston—so close, Johnston says, they can wave to one another when the foliage is off the trees. Wave, maybe, but they still wouldn't see eye to eye, at least not on the future for genetic engineering.

Collins sees ruin for a lot of companies in the near future, which will lead to hard times for all of them and a slowing of the technological transfer—if only a temporary one. He foresees the kind of collapse suffered by EDP and for the same reasons: failure to follow the rules of "rational economics." He calls the setting up of an industry and even public stock sale for some companies without "a steady and predictable stream of revenues" completely irrational.

Collins is not alone in his pessimism. Scott King saw a short, bleak future for most of the one hundred or so companies he watched regularly, although his pessimism has eased as time has passed. He found that the "circus atmosphere and overheated investment market" of the summer of 1981 seems to have settled down. The rush of new companies after limited and dwindling financing has slowed. Some companies were unable to get start-up money—and a good thing in most cases, King thinks.

First Boston's Joyce Albers believes some poorly organized small companies soon will use up their financing and return for more. They will run head-on into other "start-ups" trying for their first-round financing, and the cupboard will be bare. Companies will begin folding, others will shrivel up into pure "survival" states.

King and Albers attribute many of the problems to complete lack of understanding between scientists and their business partners. King, a graduate in enzymology, says venture capitalists enter professors' offices, "and the profs come out glassy-eyed. They've got equity, stock options!" And even Collins, a former professor of mathematics, laments of the scientists: "These fellas are coming pure and clean into this dirty world, and they're ripe for exploitation."

Maybe so, but most of the scientists involved in the field express many of the same reservations and foresee a major shake-out in the industry that will "do in" the poorly run companies and leave the good ones shaken but ready to carry out their mission of transformation.

"Shake-out" is a frequently heard word among scientists and executives, each trying to guess when that period will come when the easy money stops and the demands to produce overtake eagerness to invest in the newest game in town. Company managers speak of projects and contracts that might carry them through the shake-out and worry about poor operations that color the public perception of the entire industry.

The *Wall Street Journal*[1] ran a story about one firm that consisted of two well-qualified scientists and a board of directors that included a disbarred lawyer and another man with suspected organized crime ties. As a result of its story and other publicity, the *Journal* reported two weeks later, the scientists bought out their suspect partners.

Venturist Lodge said bluntly that he fears the credibility of the nascent industry is being threatened by "some real flimflam proposals I've seen, with some major investment houses backing them."

One scientist who thinks few firms will survive also believes it is futile to try to predict which ones will make it, because of the "singular" nature of scientific discovery. "If I were going to invest in biotechnology," said Richard Losick at his Harvard laboratory, "first I'd want to be very young—able to wait twenty or thirty years until something happens. Secondly, I'd want to spread my investments to as many firms as possible, because the least likely is the one where one wizard is going to discover 'the big one.' It'll be a company nobody's heard of, a guy with a couple of buckets, working out of his garage."

Losick's wife, Janice Pero, disagrees on the scale of that future success story. "It takes a certain amount of capital investment to do rDNA technology industrially." But she agrees that the future of genetic engineering is unpredictable and that "the big moment" is not likely at hand.

Even entrepreneur Johnston, who survived the lean years with Genex, agrees that too many companies are scrabbling for available money and many will fail. Where Collins and King stand apart from many in the industry is in the degree of failure they see ahead. Nevertheless, even they see an astonishing future for biotechnology after the storm is weathered.

Some who predict an economic shake-out say such winnowings

are a normal part of American business. Genex's Leslie Glick
noted that fifty years ago there were dozens of U.S. auto makers;
now there are three. In the 1960s beer brewers were going out of
business at a "stunning rate" even though the national appetite for
beer was going up. Fewer survivors were divvying up the larger
profits. Perkins commented that such a winnowing of failed companies might be normal, "but it's *not* a good way to make money."

Scott King believes that virtually everyone has an unrealistic
expectation of when biotechnology will deliver products in the
quantity and at the cost that they will become competitive with
existing manufacture. King thinks such optimistic forecasts as the
Office of Technology Assessment report rely overly on the views
of Glick and others in the industry with axes to grind. In his
opinion, it will be fifteen years before the "genetic revolution"
reaches the marketplace.

Collins, in his yard, directs a piercing glance into the starry sky:
"Ah, the pioneer. You know who he is? He's the guy way out in
front of everyone else—face down in the mud with an arrow in
his back."

An adage often quoted by those pondering the genetic engineering business, but Collins uses it to make a point: high technology,
the business of the new, has special risks others don't. "There is a
very narrow slice of time in which you can be profitable with
something new," he says. "If you're too early, you've spent the
money and developed the product, but there's no market—no
positive stream of cash. If you've come in just a bit too late, the
first *successful* guy there has developed the market and you find
he determines it. He sets your profit margin, in other words."

Only in that narrow "right time" slice is rational investment in
a pioneering effort possible, Collins believes, and the key word is
"rational": "Some people say the world is unfair. But in a strange
way I've come to think it is fair. Rational economics rules the
world, and those who violate rational economics pay the price—
but they often come to think the world is unfair."

The first major error Collins believes the genetic engineering
firms have made is in building their financing structures wrong,
most importantly by progressively overvaluing themselves. "Why
do people invest? To make money. There is only one rational

reason to invest in something. You see a future stream of cash, and you want some. That's it." But, he believes, the biotech companies have no steady stream of revenues that would justify their valuations, a slip made possible because money has been so easy. The prospect of fast riches is dangled before investors. "Greed takes over and fear goes away."

But until companies reach large-scale revenues—in the $25 million to $50 million range—they have only their positive cash flow to attract rational investment. Genetic engineering isn't at that point. "Some [genetic engineering] companies are valued at $40 million. Do you know what kind of cash flow it takes to make investing in a $40 million company with assets of $10 million profitable for the investor? Enormous."

Therefore, the majority of firms, which have set themselves up to do contract work for major pharmaceutical or chemical houses, have tied themselves to small cash flows, with only the rare hope of a "big strike." They offer no opportunity for what Collins calls "explosive growth."

And even more importantly to him, "No company should go public until it has a steady stream of quarterly earnings increases to offer."

Collins takes his precepts directly from EDP. In the late 1960s, he says, "All you had to do was say you were in computers and the money flowed. This was not rational investment, this was not rational economics. By the early 1970s, you couldn't sell a share of computer stock. Companies needed money for future activities —it wasn't there. Most of them had developed very bad habits, they weren't used to make a little go a long way. They're gone."

The immediate cause of the collapse, Collins believes, was a failure to produce a rapid flow of regular earnings. "Somebody gets the idea you're not sound. The idea spreads—to good companies, bad companies, any company that can't show that stream of earnings—*and none of them can.* Soon a portfolio manager looks like an idiot if he hasn't dumped the stock—and *that* is the opposite of being hot."

Some of biotechnology's majors, like Genentech, have enough money to survive a falling out with the public until that steady stream of cash comes in, Collins believes. Nevertheless, "I predict that soon you won't get venture capital to invest in this business that venture capital started."

Private companies stand a better chance, Collins says; major corporations such as drug and chemical firms can make something akin to rational investments, because they stand to benefit in a major way from genetic engineering developments. For example, any major drug company might put up large sums of money because "my ox will be gored" by the new technology, or because such investment represents legitimate corporate development activity that the firm doesn't want to undertake in-house. And finally there is the "crapshoot" philosophy, a chance to speculate legitimately, but Collins believes such investors are good for only "one round."

Collins is no parvenu in high technology. He holds masters and Ph.D. degrees in chemical engineering, a second masters in mathematics, and spent seven years teaching at Princeton, falling into electronic data processing almost by accident when he became involved with the university's computer center.

And if his view of the near future for genefacture is as dark and "bearish" as possible, he remains strikingly optimistic about the eventual impacts. Without missing a beat, Collins enthuses: "Genetic engineering has the capacity to bring about a more profound change in our society than we have ever seen before, one that enters not only industry, but strikes at the moral and religious aspects of our lives as well. We will be toying with immortality some day, no doubt of it. I don't know if we can deal with it, but it will be there. We're talking about a fundamental understanding of the life process. We're talking about a dramatic potential for good and evil."

More than from nuclear physics or computer science? "I think so, yes."

TWO COMPANIES

> There's something strange going on here. The scientists are making money and the companies are publishing.
>
> —Wife of a Scientist

Heading up toward San Francisco on Route 101 with the sky full of clouds but the autumn air remarkably clear, you turn off the flat peninsula into South San Francisco's heavily industrial

Point San Bruno, along the edge of the Bay. There, in a low-slung, sprawling modern building sits Genentech, wearing its corporate name-tag modestly low along the shrubbery. Within its single building work hundreds of scientists, technicians, executives, secretaries. Behind the reception area, closed off by display glass, sits the corporate badge of identity: a gleaming 750-gallon fermentation tank.

Fred Middleton, one of the earliest Genentech employees and its financial vice president, explains, "Until 1977 the company really was Bob [Swanson, the president], and most of our research work was contracted to universities. We've changed that entirely now. We do our research on-staff. We need the devotion of full-time people."

But no dates in Genentech's short history have the importance of the day it went public. October 14, 1980, marked a turning point in genetic engineering as industry, a milestone to compare with Cohen and Boyer's collaboration in the science, and one that illustrates a major feature of public money: it carries with it publicity, which brings new industries out of the purview of a few to the scrutiny, interest, conversation of many—for better or worse.

Genentech raised $35 million in a few hours in return for only 13 percent of its equity as its offering sold out, then was traded wildly for the remainder of the day. Offered at $35 a share, the stock quickly rose to $89 before closing just below that. At the time it was the largest single public stock offering in history.

Genentech was the darling of Wall Street, the subject of countless newspaper and magazine cover stories and of a *Time* cover[2] less than six months later.

That boom represented the sprouting of a technology from pure science, and never before had the push from laboratory occurred in such record time. In no other case has an entire cutting-edge science been translated into technology in a few years. Not surprisingly, such an eruption can bring financial headaches and even ruin for some who misunderstand what is happening, be they investors or scientists. But the speed of the transfer is also bringing the promise of genetic engineering to fruition faster than anyone imagined just a few years ago. That occurred in part because a few business executives and scientists virtually reversed roles, the scientists learning the language of the business plan, earnings per share, profit-loss, and the business executives putting money into pure research—always considered a high-risk gamble.

Of the early days, Perkins says, "When we founded Genentech, we were backing pure research. At that time, that's all it was. I'm not saying genetic engineering wouldn't have started without us. It would have, of course. But I don't think anyone else would have given it the push we did." As we'll see, that push didn't go quite the way Perkins had planned.

If after leaving Genentech's headquarters, you continue driving up U.S. 101 over the hills and past the rooftops of San Francisco to the Oakland Bay Bridge for Berkeley, you find Cetus's corporate offices, where there is a whole different manifestation of rapid corporate development. The headquarters does not appear much different, but some distance away near the North Bay waterfront is a huge old brick warehouse with a factory across the dusty street. Conversations here take place over the din of construction machinery. This could be a movie depicting industrial activity, but it is new lab space for Cetus. Cetus was expanding phenomenally on the crest of *its* public stock offering, which came six months after Genentech's, bringing Cetus $125 million to supplant its rival with history's largest stock offering.

Cetus and Genentech, the oldest and most established of the publicly held genetic engineering firms, sit like rival siblings on opposite sides of the Bay City. Cetus has exploratory fingers in every aspect of biotechnology, while Genentech has culled its projects carefully for specific goals. Cetus spreads all over the map, making major deals with many large corporations in return for large chunks of its equity, renovating buildings in various parts of the North Bay area, putting up pilot plants in the Midwest. Genentech remains quietly rooted in the San Bruno area of South San Francisco, and says it has retained better control over its equity than the other public companies.

The difference in strategies goes back to corporate beginnings, as related on that autumn day by Peter Farley, then Cetus's president and chief operating officer, later co-founder Ronald Cape's board chairman and chief executive officer. From the outset in 1972, Farley related, Cetus had in mind exploring the budding technology of genetic engineering along with two other areas it was pursuing—medical instrumentation (since dropped), and using its own screening process to develop better microbes for anti-biotic manufacture. Farley recalled that soon after the company began, Nobelist Joshua Lederberg, a science board member, alerted the firm to "very interesting work that a biologist named Stan Cohen was doing over at Stanford."

From that beginning, Farley says, genetic engineering was one of Cetus's corporate goals—yet the "myth" persists that it was a johnny-come-lately in the field, after Genentech. At that time, Cetus was partly owned by a pension fund. Then came a stroke of fate, via the U.S. government. The Labor Department decreed that pension funds could not be used to invest in speculative ventures, and Cetus's investors had to sell. The buyer: Kleiner, Perkins, Caufield, and Byers.

In this version of genesis, after a time it became apparent to Robert Swanson, assigned to handle the Cetus account, that he would learn nothing of the inner sanctum, so Kleiner-Perkins sold its Cetus shares and went on to found Genentech. It happened that Kleiner-Perkins sold those shares to a Canadian firm represented by two young men interested in biology and foreign enterprise: Inco. Seeing Cetus's light, Inco went on to put up the major seed money for Biogen.

Farley's story points up two interesting facts about this nascent but growing industry: just like the science on which it is based, the business was started by a small group of people generally acquainted with one another, and there is no love between some of the pioneers.

For his part, Thomas Perkins, ever soft-spoken, says Farley is right that Cetus opened his firm's eyes to the field, and that probably happened because Swanson was assigned to watch it—but Kleiner-Perkins sold its shares in *frustration* that Cetus was sitting on the technology waiting for it to develop, instead of pushing it toward its birth.

The drive to bring genetic engineering into bloom in the market-place has had its down side. *Public money is volatile.* No one understands that better than Perkins. In all the hoopla over the Genentech sale, few heard him express disappointment with the way it had gone. After rising to $89, Genentech stock dropped within months to the low thirties. "That tends to make people feel like they got taken," Perkins said. "They seem to feel we reaped some benefit of that price rise." Forgotten by many was the fact that Genentech sold out its million shares at $35—the offering price—and not a penny more. After the shares sold out, demand *among traders* pushed the price unrealistically high; those trading in the stock or buying from traders made and lost all the money as it rose and then dropped. Many writers looked only at the drop from the unrealistic high in assessing the company, and the industry.

Looking back, Perkins said that deciding the terms of a public offering involves some of business's toughest decisions. Initially, directors and underwriters were going to offer at $20, then they bumped it to $35—a premium price for an over-the-counter issue —because they sensed there *was* a higher demand than initially expected. Perkins still feels the offering price was right, and Fred Middleton noted that the price since then has been in the upper 30s to 40s, meaning the "market" is pricing the company in the same range its directors did. What went wrong then?

Perkins said, "I tend to think we should have offered another 100,000 shares—just another 10 percent. That would have taken a lot of pressure off at the opening; the price wouldn't have gone so crazy." The company actually would have made more money that way, but the point Perkins returns to is: "People wouldn't have felt they'd lost money on us." Further, such private volatility indicates that "people don't really understand what we're about" in genetic engineering.

When Cetus went public six months later, by most judgments it shot for the moon. The firm offered five million shares at $30 a share. Since Cetus was selling only 35 percent of its equity for that, it meant the firm was valuing itself at $400 million, a figure one long-time biotechnology executive called "completely obscene." But the offering did not sell out, unlike Genentech's. Further, Cetus saw its opening price of $30 fall to $19 a few months later, then to a low of $11 in one year, and it has since been hovering around that point.

Scott King interprets those reviews clearly: "An offering that doesn't sell out, and the stock doesn't maintain something near the opening price? That has to be called a failed offering." Another company executive was asked if in a sense the price merely indicated that the market had adjusted to a more realistic valuation of Cetus. "Maybe so," he said. "But people bought the stock at $29 and watched it lose more than half its value. Try telling them that."

But others praised Cetus for the timing of that offering, for it accomplished for that firm alone what virtually all of them have dreamed of: survival, for at least the next few years.

Farley: "Our decision to secure the kind of money we did guarantees—and I mean guarantees—the survival of Cetus as a company." The firm netted $107.2 million on the sale, and Farley predicted that by 1985, when recombinant DNA and other biotechnologies will begin to pay handsomely, Cetus will still have $90 million left, all of it earning high interest. Just as the soaring of a company's stock in trading has no effect on the firm's treasury, so the plummeting of its shares has no impact, as long as the company never needs to return to the public for more.

"The gentlemen on Wall Street keep saying how volatile the market is, how inflation does this and recession does that," Farley said. "We are out of that market, we can concentrate on achieving the goals that will yield the products that will justify everything we've claimed."

If there is a difference in attitude between Genentech and Cetus toward its public financing, there is at least as great a difference in their relations with corporate partners.

Genentech prides itself on having sold few large blocks of its equity, all under very favorable conditions. Lubrizol, interested in diversifying away from lubricating products, paid $10 million for one million shares in 1979, then bought out Inco's original shares for another $10 million a year later to give it just over 20 percent of the firm, Middleton said. Later, Fluor Corporation bought about 4 percent for $9 million, but in return brought valuable knowledge to the partnership as a major builder of industrial fermentation plants. Genentech strives to epitomize the well-run firm, and certainly among all the entrants in the race it gets the highest marks from industry analysts, who especially noted its ability to attract top scientists and—most significantly—scientists who continue to publish in their fields, an unusual freedom for corporate genetic engineers. They also praise Genentech's strong management and clear goals and note the satisfaction of clients with its products.

Cetus, bigger and brasher as represented by Farley, does not generally get as high marks for management, but is top-rated for its financing and is also ranked very high for the quality of its scientific talent. Senior scientist Dr. Shing Chang, for example, is considered one of the country's leading experts on *B. subtilis*. Noted plant geneticist Winston Brill, quoted in Chapter 1, works for Cetus at the firm's pilot alcohol production plant in Madison, Wisconsin. Cetus is criticized for spreading itself too thin, but Farley said that directors have established priorities for each of the many projects the company has investigated and, when the time becomes right, will seek "strong partners" for successive joint ventures. Thus, in partnership with National Distillers, Cetus developed what Farley described as the word's first continuous alcohol fermentation pilot plant (it does not use recombinant DNA techniques). Farley says that with its enormous financing and "forward integration strategy" toward these joint ventures, Cetus represents "the largest critical mass of biotechnology in the world."

The strategy has cost heavily in equity, even as it spread Cetus plants and spinoff-company buildings from the Bay Area to the Midwest. Cetus ran until 1977 on $5 million in venture capital, according to Farley. Then Standard Oil of Indiana (Amoco) came in with $10 million and got 22 percent of the company in return. Cetus subsequently formed partnerships with Standard Oil of California and National Distillers, and these three largest shareholders now own 65 percent of the firm.

At times, genetic engineering seems foolproof. For example, in an interview two years ago Farley pointed to Cetus's major project to develop pure fructose sweetener, and the numbers he forecast are both dazzling and important to industrial America. Within the past decade, high fructose corn syrup (HFCS), which, as noted, is not sweeter than table sugar but cheaper, has captured 30 percent of the industrial sweetener market, he said. That market is worth $11 billion a year, so that displacement means $3.3 billion a year to those providing the feedstock—the wet-millers of corn—and it means a shift away from sugar beets and cane. More significantly, corn is a largely American commodity, while most beets and cane are imported, so the shift represents a major change toward domestic producers.

That shift was accomplished through the use of the enzyme glucose isomerase.

"A major industrial dislocation took place over eight years, but you never hear word of it," Farley said, "and it's certainly not credited to biotechnology."

Farley proposed going HFCS one better, selling pure fructose, genetically engineered via Cetus-patented microbes. Company forecasters saw a market for supersweet fructose as $6 billion a year. "If we got only 10 percent of that market, I think I'd be happy," Farley said.

While genetically engineering pure fructose is still a lively dream and may well have been accomplished by the time you read this, Cetus subsequently had to pull back in its drive to do it first. In September 1982, the search was among several massive projects cut by the company, which also laid off forty employees, the first layoffs in its history. The assessment: that despite its huge capital reserves, Cetus needed to focus on fewer projects that could bring revenues sooner and not put so much effort into long-term searches for those "big boomers" that could fill its coffers indefinitely.

Soon after, yet another chapter was added to the corporate book of genesis. Robert Fildes took over as Cetus's president and chief executive officer and Farley became vice-chairman. Fildes moved from Biogen, where he had been president, soon after Gilbert left Harvard to take full-time control of that company. *The New York Times* of May 10, 1983, reported that Farley announced his resignation the previous day to undertake a new venture, marketing a specialized software package that would improve health care evaluation. With Fildes continuing as president, Farley left Cetus in "good strong operating hands." "It's off to the races again," Farley added.

The genetic engineering industry presents contradictory images: bricks and mortar and the din of heavy construction equipment, sleek corporate offices, and well-run labs headed by some of the best minds in science, and on the other side the caveats by Collins, King, Albers, and others that there is little substance here, for substance in industry must be measured in earnings.

It does seem that this industry has been created in *anticipation* of a boom—in anticipation of products with demands they would satisfy through sales, leading to profits—and that would be unusual, to say the least. These visits to Genentech and Cetus took place before any of their own products were on the market, though Genentech had delivered contract products that were well-received. Still, nearly $1 billion has been invested in a variety of companies and four "majors" remain—Genentech, Cetus, Genex, and Biogen. How did this happen? Is the enterprise doomed to fall in a bust or bound to succeed?

Analyzing the character and "weight" of the new industry through income plans, revenues, and real earnings, as analysts do, is an important way of reading the future, but Perkins, Farley, and other industry executives feel that in many ways the industry is not being weighed properly as a high-technology enterprise. Such firms must be seen as people-intensive. The inability to judge what people will do in a new technology ought to lead to a certain forbearance: there is, they say, a need to withhold judgment and to keep working.

The first goal of the industry is to replace existing technology: for that, no new markets need be created—the firms need to link up with companies that have created markets, and they have done so.

Perkins: "The best way to look at this industry is to look at the potential market. Consider the possibilities for that market. Then ask: Can Genentech have an impact on that market? How much? If you look at the *whole field* like that, and you start adding up the potential profits, you get a very big number—a *very* big number."

For example, Perkins said, drugs have a very high profit margin if the developing company hits. "We believe we have a good chance of hitting." But this is a risk industry. "It isn't an area where you can know you're going to make this much money on this date. It's not a Southern Pacific bond issue, but there is business going on here, and there will be earnings per share some day that will be used to evaluate what we're doing."

The profit *potential* also seems much larger for genetic engineering than for other new industries because virtually every industrial operation has an analog carried out via enzymes, and that may account for how a small industry was able to form in anticipation of products. If only a small percentage of the possibilities hit, the industry will be enormous, as will its impact on the average citizen.

Most companies' strategies involve bringing in enough revenue to survive until "the big one." Most do contract work for clients—as with Genentech's developing insulin for Lilly, or Biogen's similar project for Novo. Contract work does not, as Collins noted, provide for explosive growth or justify large valuations, but it isn't usually intended to; it's intended to keep the firm afloat. One scientist said appreciatively, "Genentech has just about covered its research costs with contract work." That is important because it means the better genetic engineering firms may not need to return either to public or private financing sources until they have a specific project they want funded.

Perkins: "There really shouldn't be any 'second-round financing' unless the company can come in and say, well, we *do* have a winner and we want money to produce it." And at such a point, the company will be offering what Collins says good firms must: a future but visible stream of cash.

Lynn Klotz of BioTechnica in Cambridge points to a novel revenue-producing method not usually available to high-tech companies—selling technology. "Say a big drug company genetically engineered microbes to produce a certain drug. You can sell them those microbes on contract for a certain small royalty [usually 10 percent]. But you *could* just teach them the whole cutting-edge science so they could produce their own. You could get a lot more money for that, because you'd put them way ahead of their competition."

Disadvantage: "You've taught them all you know. You've sold all you *have* to sell in a particular line. But the point is, some day they're going to work it out anyway. The small companies with the top scientists in the country have a real advantage right now, but it won't last Eventually, with that much money to be made, the big corporations are going to learn all we know. So we sell them something today that's very valuable to them, putting them ahead of their competition, and we sell it while we *can* sell it."

Nevertheless, while such strategy brings in radically more revenue than contract microbe-engineering, it isn't the big hit either, and should not be confused with it. Klotz likens the strategy needed for genefacture firms to that used by serious poker players:

"The start-up money is the ante that gets you into the game. You shouldn't think of it as anything more than that. Even if it's in multiple millions of dollars, it just allows you to sit down and play. All the rest of the income represents the small hands you win while you're waiting, the hands that enable you to stay in the game. But you're waiting for the big play, and the trick is to be able to last; the odds say the longer you last, the better your chances of its coming. You have to *be* there when the big one comes, then you have to play it right. The major discovery—that billion-dollar bug that, say, gets 10 percent more oil out of the ground—that's what it's about. That's why you're in there."

There: on the cutting edge. Dr. Morris Bell thinks most evaluations of genetic engineering have completely missed because they ignore the nature of the territory.* In a real sense, he is Collins's opposite number: "brought up on Wall Street," where his father has spent his life, he majored in Marxist economics, then "hit the Street" himself for several years. But he had always liked science, so he went back to school and became a molecular biologist. He now specializes in bacterial molecular biology at The Rockefeller University in Manhattan, and he also serves on the business board of a genetic engineering firm.

First, let's assess the financial risk: "Are we comparing a speculative new business to a maker of widgets? You can't figure the two the same. How much does a corporate jet cost? Four to five million. That's how much big companies are sinking into any given genetic engineering firm, and they're getting a lot more for their money than a new jet. How much does a Broadway play cost to produce? Four to five million, and the odds on success are even worse, but you've got loads of people who just love to invest in 'em. Of course, the point is that if you want immediate cash profits, don't invest in biotechnology, invest in widgets."

Overvaluation? "Not at all, there's a real logic to most of the valuations. First, you need $5 million to $6 million to start up one of these companies; it's not wise to start with less. If all your start-up money goes to the investors [i.e., you give a lot of equity

---

* Bell is a pseudonym for a scientist with unusual business expertise, who did not want to be quoted by name. He is not quoted elsewhere in this book.

for a given price], there's no equity left for the scientists. If you didn't need to give the scientists equity to get them, the big drug companies would have gotten them long ago." Once the equity is parceled out and some is reserved for future science-hires, Bell says, the valuations frequently work themselves out by simple mathematics into the $40 million range.

BioTechnica's chief executive, John Hunt, points out one of the unique elements of the scientists' positions in the firms: "In a real sense, there are no genetic engineers. There are engineers in every other science, but here the work is carried out entirely by the top people in the field." That creates special management demands. "You're talking about a collection of very creative, brilliant people, and far more than even in a university, they must be able to work together."

Bell: "That's right, but an engineering class of scientists will develop here." And there will be another kind of shaking out: "I think there are already some of the top scientists in the business who'd rather be back in the lab, and at some point they'll move that way as others move out of the universities."

Walter Gilbert similarly believes that soon pure recombinant DNA research will split from applied research, as has happened in every other science field, and the two branches will naturally develop in different ways.

Further, because the firms are so totally defined by the talents of their scientists, they represent a real resource *before* they've sold a product to other industries that support them. "The sources of information cost hundreds of millions of dollars to develop in universities," Bell said. "Making an investment in one of these firms is like plugging in to a terminal of the world's greatest computer. They're tapping into an incredible pool of knowledge. That's not true in the electronics industry—there is no such pool. If you want to know [computer] chips, you don't find a university scientist, you talk to the chip man at ABC computer. There's no precedent for this."

Bell, in fact, thinks gene-splicing companies should show they are taking some real risks *before* they become worthy investments. He pointed to one biotech support firm admired by King because of its traditional steady sale of quality products to other genetic engineering companies. "And that's exactly why it may always be a good company, but it will never make a huge hit," Bell said. "It will never go right out there to the edge and take a bunch of big risks on projects, some of which will never pay off."

But back to the widgets: "Biotechnology doesn't even come off badly by comparison. Let's say you're into widgets and you're optimistic, you see a 10 to 15 percent increase per year in demand, and you're well-positioned among other companies in widget manufacture. If that were the case, you'd expect to get fifteen to twenty times earnings for your stock. Now if you can see a cash flow of $30 million to $40 million you'd say that justified an investment of $600 million to $800 million." Looking at the whole genetic engineering industry, Bell says, that's right on target.

"Look at just one product: insulin. Worldwide market, $500 million. Genetic engineering firms get about 10 percent [royalties], so if you've got the strain to make it that's $40 million a year. And that in itself would support a valuation of your company of $400 million."

Bell points up a key distinction those practicing in the field say is not being made by most analysts. The near-term future of the *industry* is not bad, even though many individual companies may go down. "A lot of them will deserve to," Bell says. "When it comes to second-round financing, they'll either be very good, with a real 'score' they're hoping to finance, or very bad, with no products."

Nevertheless, Perkins is apprehensive about what he calls the "bandwagon financing" of the companies in the field. "When you're the first out there, getting money isn't so bad. I'd hate to be fifty-seventh, and now there are at least 100." Firms that can't show patents and products indeed will fold, Perkins feels, but the backlash "will hurt the good firms, like Genentech, too."

While many see growing threats in the field from foreign competition, Bell does not. The Europeans have tried to create their own industry with government sponsorship; but because that has meant no equity to scientists, they have had little luck. Biogen, for example, is a Netherlands registered company but operates out of Switzerland and Cambridge, Massachusetts, and does provide its top people with equity. It has no government ties, and Gilbert is its chief executive officer. As Gilbert noted, Biogen was created on the American model, with a strong emphasis on control by its scientists.

The Japanese, Bell says, "are extremely good at putting established knowledge to work. It's very boring, dull, and of great commercial value. But we are not of the 'we'll make it work better' school. We break new ground, always have. That's our national personality and it's perfectly suited to this industry."

Others are more concerned about the Japanese competition, and they point to their heavy investment in U.S. genetic engineering firms and their eagerness to buy technology—to pay handsomely in order to be taught the most advanced science. Further, if achievements in molecular biology are not on the whole as advanced as those in American universities, the Japanese have long been the acknowledged world masters in fermentation technology, and they have committed themselves to developing a genetic engineering industry as a national goal.

But such competition does not trouble Bell. He says it only further underscores the role of biotechnology in America today. "Don't think of all these genetic engineering firms as a bunch of separate companies if you want to figure out what's going on here. Think of them as the high-tech research and development arm of U.S. industry."

FINDING YOUR WAY
IN THE GENE AGE

We have made a lot of points about science and the business of genetic engineering that may be useful in evaluating products and companies, for large and small investors or for research directors. We'll conclude by summarizing them.

Who Are the Company's Scientists?

Genetic engineering is extremely difficult in almost all cases where it's useful. If that seems simplistic, it is a fact that leads to important conclusions. To put a simple gene into *E. coli* and just get expression is not difficult. To maximize gene expression and overproduce protein is very tough. Not a lot of people have the bag of tricks needed to do the job.

The cDNA cloning technique needed to get genes for rare proteins from animal and plant cells is a tedious one and few scientists are good at it. How can you tell if a company has good scientists? The company's track record is the best single indicator. For example, Biogen and Genentech already have completed several genetic engineering projects, carrying them from the laboratory bench through, in some cases, industrial scale-up. Some of these projects, like interferon production, involve extremely rare genes. At least that is assurance that these companies have the technical capability of staying at the forefront.

Newer genetic engineering companies—which means most companies in this industry—haven't established records yet. But there's still a way to get a lead on their chances, by directly investigating their scientific staffs through information from a library or the company. What kind of publications have their leading scientists got to their credit? Are they or were they pioneers in university labs? Past pioneers tend to be future pioneers. Most important, if a firm's top scientists don't have a number of publications centered on the major aspects of genetic engineering, that's a strong indication they can't carry out the procedures.

Most companies boast advisory staffs of molecular biologists—well-known ones wherever possible—but in evaluating the staffs make sure they *also* have full-timers experienced in genetic engineering. No matter what abilities the advisory staffs provide, it is the full-timers who will directly affect the quantities of proteins produced and the time scale to production.

Pioneering is important in another respect. Developing new techniques puts a firm in a good patent position, and licensing of patents may turn out to be a large source of future income—much larger than for universities, for example, which do not seek high returns on licenses. Patent position may be especially important related to organisms other than *E. coli*, and to fermentation scale-up. The newness of biotechnology patents certainly clouds their future. Nevertheless, the companies with the best scientists and connections with the best fermentation engineers stand the best chance to get projects into commercial production and get patents.

What's Their Line?

Proposed product line is important no matter what the company size. A small firm should have at least two and perhaps more products planned that will offer *short-term* profit potential. Southern Biomedical, which recently went out of business, relied heavily on classically produced interferon for income. One product is not enough; there is no certainty that any given product can be made in enough quantity to be commercially useful. Sometimes the best genetic engineer will be helpless to turn around a failed project; if so, there had better be another in the works.

Most companies will have to begin getting income by the end of their first two years, because research and development is very expensive. Genetic engineering R&D is less costly than that in, say, nuclear physics, but it isn't cheap. Most people in the industry say it costs more than $100,000 a year to support one scientist's work. That means a company of only twenty scientists could use up to $2 million a year. Most new firms' first-round financing is in the $1 million to $5 million range, and that means only a few years' grace at best before they either have to go "back to the well" or start bringing in money. The well may be dry—certainly a more likely possibility for those who can't demonstrate achievement.

This may have been a major problem for IPRI. Frequently touted as a good, up-and-coming company, it ran into serious financial difficulties in 1982. The firm employed more than 100 scientists, however—meaning it needed contracts to cover a payroll and overhead expenses of over $10 million a year.

The product line on which the company is toiling is also important because in two years very few genetically engineered products will reach the production stage. The smaller companies will have to make money by selling engineered organisms to larger companies, among other ways.

Potential near-term revenue producers include proteins with pharmaceutical uses (for example, human insulin), and such proven industrial enzymes as glucose isomerase, already widely used to make high fructose corn syrup, and α-amylase, the starch-degrader used to make industrial alcohol and other chemicals. It's easier to draw a potential-profit-margin picture for these products because they have existing markets with known economic patterns. By contrast, even though we believe interferons ultimately will prove important in treating viral diseases, that isn't certain. FDA drug testing requirements add another layer of uncertainty to the economics of making these proteins.

Companies that plan to get most of their revenue from proprietary work—done on their own, not under contract—especially ought to have a well worked-out profit plan.

Does Somebody Up There Like Them?

Does the firm have one or more contracts with large companies? Contracts can be an important source of short-term revenues for a small company, but more importantly, they indicate that somebody at a high industrial level has studied the company and believes it has the skills to handle the job. However, there are other key questions here. Does contract income cover expenses? Some companies expand too fast, so they spend far more than they can reasonably expect to take in on contracts. Ordinarily, nothing succeeds like success, but this is a way to fail by succeeding too well, especially if other income sources aren't at hand. Further, it stands to reason that a company with a fairly conservative bent that demonstrates it can live within its income will more likely be around after the virtually inevitable shake-out.

What's Down the Road?

Survival is nice, but overwhelming success is nicer. Several long-term projects in genetic engineering have the potential of being billion-dollar winners—if they can be shown to be economically feasible. Such projects are found in the areas of nitrogen fixation in plants, energy production from waste biomass, and basic feedstock manufacture for the chemical industry. Forecasting the economics for these projects is torture. For all these products, the economics depend on the future price and availability of petroleum, and for some there is a question of whether they can be carried out at all.

One must be wary of a company that speaks of its future only in terms of these high-risk projects. But any forward-looking firm ought to have a few, far-out "great notions" in its project bag, to keep alive the chance to become one of those that will be remembered for opening up the Gene Age.

Chimeras

Chimera (kī-mē'-ra): Head of lion, body of goat, tail of dragon. Fire-breather. The original chimera was a whimsical piece of work, whether of nature or of a fertile Greek imagination. But there was no whimsy in Patent No. 4,237,224 as it issued December 2, 1980, announcing itself simply as:

PROCESS FOR PRODUCING
BIOLOGICALLY FUNCTIONAL
MOLECULAR CHIMERAS

The Stanley Cohen–Herbert Boyer process patent details how to make a "plasmid chimera," so named because of the unnatural relationship of its parts: plasmid DNA conjoined with the gene coding for human insulin, stitched into a ring, and neatly inserted into an *E. coli* bacterium, thus creating a novel life-form.

In August 1982, the patent office turned down Stanford's application on behalf of Cohen and Boyer for a patent on the microbe itself, but Stanford officials insisted the problem was a technical one that would be resolved.

Patents are grants for invention. They grant temporary ownership rights to those who use scientific facts to twist natural elements into new shapes, press them to new functions; they presume there are many new things under the sun. Discoveries of principles and natural laws are just that: the elucidation of something that has always been there, regardless of how previously unknown, some part of all that is presumably eternal under the sun.

That boundary between the novel and the natural, often difficult to discern, was the site of the major legal battle so far in genetics, one that ended in 1980, when the U.S. Supreme Court decided that whether an invention was alive or not played no role in its patentability. The decision involved not the critical Cohen–Boyer discoveries that put recombinant DNA "on the map," but a relatively obscure microbe created by General Electric scientist Ananda M. Chakrabarty.

Different natural varieties of microbes are known for their ability to digest various simple hydrocarbons, but none can eat and transform the complex variety that make up crude oil. Chakrabarty used a combination of naturally occurring plasmids from different strains of microbes, an older process that is only genetic engineering in the broad sense, to develop a strain that could eat the whole thing. Even though neither the scientist nor GE believes the microbe is ready for commercial development, the Court's grant of "ownership" of this microscopic living thing, by a 5–4 ruling, marked the watershed. That was June 16, 1980, but long before the Court clarified the issue, much controversy and no little confusion had been injected into the dispute.

Some writers had been wondering if the patentability of life suggested terrible new forms of slavery in which cloned humans would march about with patent numbers holding them in bondage down through their generations. Others suggested that science had been given new license to create monsters and conjured visions that Chief Justice Warren Burger dismissed as "a gruesome parade of horribles." And there were still others who worried that the profitability of patented microbes would lead to secrecy in the world's leading biological laboratories, a trading of scientific progress for royalties.

Unraveling the implications of the Chakrabarty decision requires first coming to grips with what the ruling did *not* do. Patent lawyers, who fought for the ruling as a friend of the court, generally are agreed that the importance has been greatly overblown. To begin with, a patent was granted on a living organism to Louis Pasteur in 1872 for a strain of *naturally occurring* yeast, and that may not even have been the first such grant but one that stands out because of Pasteur's fame.

Bruce Collins, an attorney who wrote for the American Patent Law Association in its filing as friend of the court,[1] explains: "The Patent Office then probably treated the yeast as simply an object of commerce and did not specifically think of it as a living thing." By the middle of the twentieth century, microorganisms were described as part of a large number of patent applications involving industrial processes, Collins pointed out—most notably fermentation processes for the production of antibiotics and vitamins. But in those cases the invention involved a new *method of using* the microorganism, not the microorganism itself.

Patents on different kinds of inventions are available: process patents may be obtained on new methods used to make something old or new, and product patents are given on the "new things" themselves. Process patents involving microbes as integral parts had always been available. Product patents had not been—but neither were they needed. No new living things had been created. Or had they? What about hybrid plants? Addressing just that narrower issue in 1930, Congress added plants—very much living things—to the list of novel material that may be patented. The Plant Patent Act made no reference to animals or to microorganisms. One question Chakrabarty brought before the Supreme Court was whether such protection ought to be given to *any* novel living strain. Objectors said that had Congress intended non-plant life to be covered, it would have so specified.

Chakrabarty's attorneys argued otherwise, and the Court majority agreed with their interpretation. When any application is made to patent an invention, a full, detailed description must be filed explaining the makeup and function of the "device." The only reason plants required specific congressional action was because providing a complete description of their structure would prove impossible, and proponents argued that the fact that the plants were alive simply had no bearing on the act.

Dissenting Justice William Brennan argued that Congress had foreseen the question of non-plant patent rights in 1930 but had not resolved it, and therefore the Court should not. The question was a narrow one.

But to Collins and others in the field, a ruling against Chakrabarty would have gone against the intent of patent law; a ruling in his favor more or less maintained the status quo. William O'Neill, a consultant to Stanford on the Cohen–Boyer patent license, said Chakrabarty "would have been news had it gone the other way. Then it would have been a complete subversion of the intent of the patent law, whereas this is just a logical extension of it."

Cetus's Peter Farley, whose firm also acted as friend of the court, said, "Our work in no way slowed in anticipation of the decision, nor did we anticipate changes in our work depending on which way it went."

Despite "exaggerated media interpretation of the results," Farley said, "In a sense the positive impact was in the Court's bringing genetic engineering as a commercial enterprise to the attention of the entire country."

The patent issued to Stanley Cohen and Herbert Boyer in December 1980 is the most interesting one in the field, and the progress of its licensing agreement, developed by Stanford University, is being watched closely by everyone in the industry. That process patent involves the prototype tools of genetic engineering: the plasmid rings of DNA and the variety of cutting and sticking enzymes Cohen and Boyer used in their 1973 experiments. It could have been issued had the Supreme Court ruled against Chakrabarty's product patent application, because process patents have issued for years. In short, with the current denial of Cohen and Boyer's product application, there is no patent resulting from the Chakrabarty decision.

But there is something very special about that Cohen–Boyer process patent. In some ways it may be the "Spirit of St. Louis" of genetic engineering—not born of the first recombinant DNA experiment, not even a patent on a "novel life-form," but a patent on the elegant bug-making process that electrified the scientific and business world to the possibilities of rDNA.

Despite the importance of the claim on the basic engineering process, patent lawyer Collins believes additional Cohen–Boyer claims concerning *production* of protein from foreign organisms will have an even greater impact. The genetic engineering process generally would be practiced just once in the course of developing a new microorganism; in addition, it might be used on a university level, where infringement charges would not be likely. But once the microorganism is made, then a firm would have to use these additional Cohen–Boyer techniques to actually produce protein, which in theory, Collins says, "covers virtually every application of rDNA at the moneymaking level."

Stanford University has high hopes for the Cohen–Boyer patent it administers, and developing and promoting the license under which it and the University of California will earn royalties was largely the work of Niels Reimers, head of the Office of Technology Licensing there, and Andrew Barnes, an associate in the office. Barnes has since left Stanford. If the patent is the Spirit of St. Louis, Barnes was the "barnstormer" who showed it off to the world. Stanford developed its licensing plan to be a model of university–industry relations. As with most university licenses, it was not designed to make a fortune, but to get maximum exposure for the patent, reduce the chance of infringement fights, and bring a modest income as efficiently as possible.

Reimers and consultant William P. O'Neill wanted the Cohen–Boyer license to bring in supplemental research money for Stanford and the University of California, but at rates low enough that all interested firms could afford to sign up rather than fight in court or evade payments. Once the terms were worked out, it was Barnes's job to sell the idea, and he traveled throughout the United States, Europe, and Japan.

Reimers explained, "This wasn't just another patent. It was the central patent to genetic engineering." Because for now virtually all rDNA work requires use of that patent, "We wanted this to be the prototype for future patent licenses." Cohen and Boyer refused all income from their discoveries, normally 50 percent of the royalties, and that enhanced the potential for the universities' income.

Barnes explained the "deal": Companies using the process pay a royalty of 1 percent of all resulting product sales, up to $5 million in sales. On sales of from $5 million to $10 million, the rates drop to ¾ of 1 percent, and over $10 million they drop to ½ of 1 percent.

In order not to put U.S. firms at a disadvantage with foreign companies that might not be paying royalties, Stanford set a royalty of only ½ of 1 percent on all foreign sales.

The minimum license fee is $10,000 a year for all companies

signed up, but with the aid of major firms in the field, O'Neill and Reimers devised an elaborate credit system to encourage early sign-ups—and that part of the plan paid off well. By December 15, 1981, the deadline for getting credits, seventy-three companies had applied for licenses, meaning a guaranteed $720,000 a year for the universities. Reimers had predicted Stanford would sign forty to fifty, while a more optimistic Barnes had expected fifty to sixty.

But those licenses could be threatened. If, contrary to Reimers's belief, the product patent is not ultimately accepted, many believe licensees will simply stop paying. At the time of the patent office turndown, Reimers said, the universities had gotten about $1.4 million in fees.

The importance of the large infusion of cash, Barnes said, was that it provided a sizable "war chest" for court challenges on the patent or action against infringers. "Initially we're going to have to take a tough stance and show we mean business," Barnes said.

Patent lawyer Bert Rowland, who filed the Cohen–Boyer application in 1974 and saw it through for six years, sees a lot more patents in the future of genetic engineering, unless the field itself inexplicably dries up. "If the whole thing goes away, there will be no patent activity, but that's the only thing that will stop it." Against those who believe infringements will make trade secrets the "protection of choice," he noted that antibiotics offer the same chances for patent theft, yet most of them have been patented and have brought large profits to the companies that developed them. Further, in most cases, fights against infringements have been successful.

It remains to be seen whether this analogy applies. Generally the *end-product* is patented on a new antibiotic, so infringement can be detected on the open shelves of the drugstore. Genetic engineering generally deals with *known* end-products, to be manufactured using newly devised methods or microbes, and they operate at the far more secret production level, where infringement might be hard to detect.

Many scientists and patent lawyers believe the genetic engineering field will be dominated by trade secrecy rather than patenting, a contrast that is very important. Patents are not entitlements to secrecy—quite the opposite. They offer inventors a limited monopoly on their work, usually seventeen years from the date of issuance. In return, the inventor must file a description of the invention so complete that anyone reasonably knowledgeable in the field could duplicate the procedure. Because plants and microorganisms can never be exhaustively described, a companion procedure many years old allowed for deposit of actual specimens in culture banks, in addition to the descriptions. Presumably a similar culture deposit will be used for product patents on novel life-forms.

A trade secret is the perfect contrast to a patent. As with the formula for Coca-Cola, reportedly known to only five people in the world, if someone figures it out, they can use it freely without limitation. If the formula were patented, anyone could learn the secret, but would have to pay royalties to use it commercially. Why use trade secrets? Because many companies believe it is easier to defend their secrecy than to defend patents against infringement.

For example, suppose a company patents a strain of *E. coli* that produces insulin better than any other. Another lab follows the procedure, obtains the strain, but classifies this process and microbe—actually stolen—as its own "trade secret." Noted one patent lawyer, "You may have a terrible time trying to prove that the strain you believe they are using is really the one you've patented." The patented strain might have to leave a telltale sign in the product for infringement to be proven.

Massive systematic infringement is especially of concern to European companies, which are afraid that patented microbes will simply be "stolen" and used in other countries in which there are no patent laws or in which discovering and proving infringement could be virtually impossible.

To Reimers, the *real* importance of patents is to prevent secrecy from coming to dominate commercial rDNA and strangling exchange of information, just as many scientists fear. And interestingly, there probably would have been no Cohen–Boyer patent had it not been for Reimers's persuasion. In 1974, with the deadline for filing application just weeks away, Cohen was set against the idea because he believed colleagues would see it as his attempt to monopolize his findings, Reimers recalled.

"The great debate over imposing a moratorium on rDNA research was reaching its height," he said. Cohen and Stanford's

Paul Berg were among the leaders in the fight for the moratorium. They believed an application at that moment would seem part of a strategy to keep control of rDNA development. "But the main thing to Stan was that he wanted the process to be used widely," Reimers recalls, "and he was afraid patenting would limit its application."

Reimers changed Cohen's mind by recounting an historical incident. Oddly enough, the story appears to be apocryphal in part, but it was true as far as the major points Reimers wanted to make:

In 1928, the story goes, Alexander Fleming published his discovery of penicillin, indicating it might have curative benefits against bacteria, but he refused to file for a patent in that spirit of sharing knowledge that Cohen now wished to embrace. However, drug companies refused to work on developing penicillin for fear that their research and development money would be wasted, because competitors would simply duplicate their results for nothing.

The result was medically disastrous. Penicillin was not used until World War II—a dozen years after it might have been available—and it was then developed commercially by an American, and made its appearance on the battlefield as a *secret* of the allies.

The source of Reimers's story was Betsy Ancker-Johnson, whom he recalled had told it to a congressional committee some years before when she was Assistant Secretary of Commerce. Mrs. Ancker-Johnson and a former aide told us in turn that they had gotten the story from a book entitled *Miracle Drug: The Inner History of Penicillin*.[2] As with any story passed through many hands, this one had lost key elements and suffered some revisions, but the major relevant points were right:

Author David Masters recounted Fleming's discovery, and noted that penicillin was a laboratory curiosity known to have some antibacterial properties but—interestingly, much like interferon—was for years impossible to produce in quantities large enough to study. A few dogged researchers picked up the puzzle one after another for twelve years, trying to learn its properties, delayed by the difficulties inherent in the investigation (not by lack of *protection*, at this stage). Then, just as Britain appeared to be losing the war with Germany, the most dogged of all the researchers, Oxford Professor Howard B. Florey, made enough of the drug for him and his colleagues to realize it was not just a scientific curiosity but a miracle cure for many deadly bacterial infections. The British scientists were determined to see it brought to commercial production, at no benefit to themselves and for the good of all mankind—and they did so, eventually.

Professor Florey desperately pressed British drug firms to make large quantities for clinical trials, but he found the fermenters bogged down in the war effort and constantly threatened by bombs.

But, author Masters pointed out, another concern may also have been operating: Many firms may have feared that as they attempted to scale-up natural penicillin for production, someone would discover a way to synthesize it and put them out of business. That had happened more than once before. As a matter of fact, a patent by Florey *would have* quieted this fear, but that does not appear to have been Florey's concern. He and a colleague flew to America and persuaded the U.S. Department of Agriculture to make the drug in large enough quantities for clinical tests. But at this point the story recounted by Masters contains a great irony apparently not known to Reimers: an American USDA scientist filed for a patent on the corn-steep liquor method he had developed for batch-fermenting penicillin. That patent was held by the USDA in this country, so Americans enjoyed the fruits of their tax-sponsored research at no cost. But that scientist, as was customary, got foreign patent rights *himself*; all the British, including Florey and Fleming, had to pay royalties to the American who had commercialized their discovery, while they got no financial reward. Fleming, Florey, and Ernest B. Chain, who also did crucial work on penicillin, won the Nobel Prize in 1945 for the difficult discovery and development, but Fleming noted the patent-rights issue with no small amount of disappointment in a speech in New York that year.

As with many such old stories, this one reverberates hauntingly through the current discussions of university–government cooperation, as we'll see in the next chapter.

Confronted by Reimers's warning on the danger to the public in his not seeking patent protection, Cohen relented, and Stanford filed one week before the expiration date in November 1974. But the road ahead was rough. Reimers recalled that Cohen was standing in a crush of scientists at the rear of a conference as the moratorium was being debated at the Asilomar Center in California, when rumor of the Stanford application spread. The conference speaker spotted Cohen and asked him to step forward and explain. Cohen told Reimers he then took "one of the longest walks in my life" to the podium.

Granted that the universities will share the income from the patents, what will even $700,000 mean to major research institutions? Little and much.

Stanford's annual research budget is $150 million, which places that private university far below the vast, public University of California system, which conducts $500 million in research every year. Roger Ditzel, UC patent administrator, noted that roughly 10 percent of all university research in the United States is conducted in a branch of the University of California. Patent income has doubled in the past few years, but none of that income is from genetic engineering. "We hope the Cohen–Boyer patent pays off, but let's say we're not spending the money before we get it."

But Ditzel also pointed out that because the function of licensing is getting products to market, UC officials determined there should be a greater commitment to that goal. "I hope our patent income develops *because* we're doing a better job of transferring technology," Ditzel said. "University investigations are not designed for the marketplace. Good old industrial sweat and toil are needed for that. We're talking about getting companies to invest a heck of a lot of risk capital in embryonic products. With a patent license, we assume they can reduce that risk a little."

Though recently tripled, the UC patent income is now only about $1 million before inventor royalties are subtracted. That is, the gross represented only 0.2 percent of the research budget. But if patent income is small, it also comes without strings attached, a boon to research administrators. Both Ditzel and Reimers point out that these unrestricted funds are invaluable in keeping departments running smoothly.

Stephen Atkinson, executive secretary of the Harvard Committee on Patents and Copyrights, put it this way: "Unrestricted dollars are simply worth more than other kinds of money." Such income can be spread around at the discretion of administrators, unlike funds for sponsored research. The latter, whether given by the government or industry, must be painstakingly accounted for. Nevertheless, Atkinson also believes that in the long run, "Patents will be less important than know-how and trade secrets, in terms of competition within the industry."

In the future, patents will be easier to defend: "As differentiation occurs in the field, applications of genetic engineering in medicine will become very distinct from those in industrial enzymology, and those distinct from energy development." Atkinson says the resulting inventions will be quite specific, lacking the nearly universal sweep of the Cohen–Boyer patents, a sweep that makes them vulnerable in court.

What, then, of the potential that patenting will set us up for the "gruesome parade of horribles" or the wholly owned cloneman?

One of Collins's coauthors noted sharply in the friend of the court brief that the specter of patent-induced slavery "is quickly laid to rest by reference to the Thirteenth Amendment" forbidding slavery. He went on to point out, "Valid property rights in living entities have been recognized as long as humans have existed, from domesticated goats and plots of Indian corn to today's vast herds of sheep, cattle and pigs" as well as "the prize bull, whose owner by virtue of a 'monopoly' and current technology earns a good profit while providing a dairy farmer with an opportunity to improve his herd."

Stanford is attempting to use its patent license to reduce the risks to human health often associated with genetic engineering— although the point may be moot now. Stanford told the National Institutes of Health, a cosponsor of Cohen and Boyer's research, that the university would enforce compliance with the NIH guidelines. The guidelines had then applied only to government-funded research, not to that done in private industry.

Now the guidelines have been considerably relaxed, a move by the NIH hotly disputed by some scientists and others watching the field, so the requirement is far less important.

Reimers contends that the risks to the public from genetic engineering probably are greater in university research than in industry, anyway. Industrial development involves scaling-up production and yields of microbes already familiar from laboratory research. It is in the laboratory that the unknown and the novel are studied. Reimers says, "It's my personal feeling that risks are much greater in pure research. It doesn't make a lot of economical sense for a commercial scientist to take such risks. A research scientist has no one looking over his shoulder."

As we'll see, there is sharp disagreement on this point and on the potential hazards, in general, on recombinant DNA technology, and just as sharp disagreement over the impact of patented chimerical life on the universities that spawned it.

Academe, Inc.

As the genetic revolution sweeps through the biology and chemistry departments of America's leading research universities, other forces are as radically altering the relations between academia, government, and business. Several changes in federal law and policy have made it possible for private firms to invest in university research to an unheard-of degree, just as the government is sharply curtailing its own research sponsorship. This opening of one door as another closes is difficult to synchronize, and as always major changes in the law raise new questions of ethics and propriety.

Congressman Albert Gore's house subcommittee is investigating new contracts between industry and universities, notably a controversial $500 million pact between Hoechst A. G. and Massachusetts General Hospital (MGH), an agreement that offers the giant German firm a direct line on the latest research of the hospital. As an affiliate of Harvard Medical School, MGH does a great deal of research, and Gore fears that, even more significantly, the agreement will give the foreign company *title* to much research that was carried out with U.S. taxpayers' money.

The problem here and in other such agreements is not a simple one. Gore says flatly, "The public policy questions we'll be confronting as a result of the advances in biotechnology are going to be more difficult than any public policy questions we have ever faced. We don't know where the advances are taking us, anymore than [Portuguese explorer] Prince Henry the Navigator knew where

technology would lead his ship. This map of life itself will probably lead to completely new worlds of which we now have no
knowledge; I think all the participants in this debate have that
sense."

What is happening here, he says, is that "the traditional lines
of separation between the institution we call the university and
the institution we call a private corporation are bleeding together
in a way that has consequences we haven't fully thought through.
Not only are they splicing genes—they're splicing institutions and
coming up with hybrid forms that raise very serious questions."

To a large extent, the emergence of biotechnology just as this
shift is occurring is pure coincidence. But for better or for worse,
a great many different events did come together to produce the
schism between the university and the genetic engineering industry.
The origin of this upheaval really lies in the conservative drive to
transfer initiative from government to private enterprise. The major
expressions of this drive in academia are massive cuts in the
National Science Foundation and National Institutes of Health
grants that have nourished American research since World War II,
passage of the Universities and Small Business Act in 1980, and
a change in the federal stance on institutional patent agreements.

It isn't necessary to delve into the intricacies of federal policy
to understand something of what is happening: we only need look
at the world before and after 1980.

Before—Since World War II, most university research has been
carried out with federal money. The aim of the federal funding
was partly to increase human knowledge, but an equally strong
motive was to spur American industry, by making new discoveries
that could be transferred to the marketplace. Thus, the government's aim has never been to make money directly on research.
Before 1980, the government held nearly all patents granted for
work done with public money. Further, if a private company contracted with a university lab that had *any* public money support-
ing it, the results of the work would almost always go to the
government. The aim here was to prevent exclusivity, because the
government offered licenses freely on patents it held. The operative philosophy: that U.S. taxpayers should not pay for the same
work twice. They already paid for the research through funding.

After—Universities now hold the patents to work done on federal contracts. The operative philosophy: universities are better able to bridge the gap between laboratory and marketplace, to accomplish the transfer of technology that is still a major aim of government funding. Even more significantly, firms now may contract with universities for exclusive rights to the results of their research, even if some federal money is involved in a given laboratory. The new requirements ban such agreements only when there is significant federal participation in a particular research project. Thus, although universities such as Stanford do not expect to make a lot of money on patent licenses per se, the changes allow for major corporate participation in research—and that does mean a lot of money. And a lot of headaches.

Is the taxpayer paying twice? Not at all, say those who support the new way, because the heavy cost in making new discoveries into reality is in development, not research.

Walter Gilbert says, "Interferon is the perfect example. The basic research in interferon biology consumed a fraction of a million dollars. The applied research required to take it through the stage of clinical testing and all sorts of other steps will take $20 million to $50 million. That's typical."

The taxpayer, in other words, never did pay the development costs, and no company *will* pay them without a guarantee of exclusive rights to the results. This point is at least partly exemplified in the later problems of penicillin development previously mentioned.

But to Gore, the Hoechst–MGH deal exemplifies many of the flaws of the new way of doing science business. Asked for an example of the "hypothetical worst case" of losing taxpayer-supported research to private and even foreign benefit, he offers:

"Let's say the U.S. government gives an institution of higher learning $25 million a year [roughly the amount of federal grants to Harvard in all areas] with the hope that money will lead to new discoveries, discoveries that will improve the lot of mankind and will allow American companies to be in the forefront of that effort. Let's say that a foreign-based company concerned about its lagging position in biotechnology wants to get access to the leading edge of new discoveries in the United States and funds a contract with the same institution . . . to give it the *exclusive* right to everything that institution develops, that makes certain, from a practical point of view that that occurs. Well, I think that raises the possibility that U.S. taxpayers will be deprived of what they thought they were getting in return for their investment."

Gore's example does not directly apply in the MGH case, because *Harvard* research and its results would not be included in any such pact between Hoechst and MGH, but he says that, in a nutshell, that is how he thinks the agreement reads. Gore says the contract gives the U.S. research hospital the legal obligation to ensure that Hoechst gets exclusive rights to everything, including collaborative work in which its money is commingled with federal funds. In fact, Gore says, the contract explicitly forbids commingling of Hoechst money with any other private capital, and expressly permits its being commingled with federal money—so long as Hoechst gets the resultant rights.

When it was pointed out that the Cohen–Boyer patent license can be bought by Hoechst or anyone else for a mere $10,000 and also represents a major investment of tax money, Gore replied, "But Hoechst can't have it exclusively. It's the exclusive ownership of a U.S. development by a foreign company that I think makes a major difference here."

At the moment there is no mechanism to distinguish a foreign from a domestic company, and the Universities and Small Business Act explicitly provides for exclusive contracts for companies on work done, *though it must be done on private money.* Nevertheless, the headaches in the new university–business relationship are hardly confined to foreign affairs.

Well before Hoechst entered into an agreement with MGH, as mentioned earlier, Harvard considered forming its own genetic engineering company. Under the plan, after forming the company the university would lend its name and reputation to the firm and remain a minority shareholder. Whether this might have been a reflection of a great new hybrid or not, the faculty protested that the university had no business in business, and the plan was dropped. Despite the failed attempt, and still urging extreme caution, Harvard President Derek Bok is optimistic about the future of corporate–university liaisons.

In a talk before the Committee for Corporate Support of Private Universities, Bok[1] said, "Radical critics talk about how such arrangements will cause universities to be co-opted, overwhelmed, and manipulated by business. The truth of the matter is that there is much less red tape, much less emphasis on effort reporting and punching time clocks when you work with business as compared with federal funding." Most of the time, "the businessmen are interested in funding the kind of work our faculty would like to do anyway."

Such assertions are hotly disputed by others in academia, as we'll see, but Bok confronts the issue as one that must be dealt with. Referring to the attempt to form the Harvard company, he said, "Harvard's problem was only one piece of a much larger effort we, together with other major universities, are engaged in —trying to see what we can do to be more active in translating scientific knowledge into new products and innovations." Such translation is part of the university's duty, he said, "for it is important that we demonstrate to taxpayers that the research they support has practical applications that will benefit them."

Stanford President Donald Kennedy,[2] a biologist, told a congressional committee that America's universities with strong research departments have developed in a partnership with government, accomplishing over two-thirds of the basic research done in the country. But he warned that "the enterprise we have built is a fragile one." He urged closer ties with industry, which now provides only 3½ percent of university research funds, but warned that risks to the free exchange of ideas should not be ignored.

"There is the prospect of significant contamination of the university's basic research enterprise by the introduction of strong commercial motivations," he said, "and potential conflicts of interest on the part of faculty members with respect to their obligations to the corporations in which they have consultancies or equity and their obligations to the university."

Gore is concerned about this possible "contamination" as it relates to universities' relations with Congress. "The public policymaker has often turned to the academic community for advice on uses of institutional agreements. I don't look forward now to sorting through disclosure statements about stock ownership and managerial positions while trying to evaluate the guidance offered."

Nevertheless, Bok and Kennedy see the three-way interchange between the university, government, and business as a major dynamic in the economy and a major requirement for what Kennedy calls "the reindustrialization of high technology." Syndicated columnist Joseph Kraft wrote recently that high technology is the key to restoring the economies of the developed nations: "New products create new jobs, and to a larger extent than commonly realized."[3]

THE BRAIN DRAIN

The torturous self-examination over the commercialization of genetic engineering hardly marks the first time academics have quarreled with "the hands that feed them." Universities began as creations of churches, were founded later by royal governments, still later by wealthy industrialists. At each stage throughout history these academies have fought with their benefactors for the right to remain honest and free in the pursuit of knowledge. Thus, if Stanford is now questioning its proper relationship to corporations, it must be remembered that Leland Stanford, Jr., provided its founding money out of his railroad empire. Does that make all this much ado about nothing? Not according to knowledgeable scientists in both academia and business.

Shing Chang, himself a senior scientist with Cetus and as noted one of the country's leading *B. subtilis* experts, fears the shift to big business may seriously damage the country's research effort if care is not taken. He cites a tragic precedent: German theoretical biology was diverted from its course by industrial chemical interests at a time when it was the best in the world, after World War I. It never recovered. "If we had done that to biology twenty years ago, we'd never have found a restriction enzyme. Nobody would have supported commercial research into something that would seem to have no immediate commercial value."

If there is a danger of research being diverted from its proper aims, there is also a danger of *researchers* abandoning basic for applied investigation. This year for the first time, Chang noted, candidates for openings at Cetus had the credentials that would have guaranteed good university positions. The big surprise: "It isn't just that they came here instead that marks such a change in trends," he said. "They never even applied to universities for jobs. They just never bothered."

Dr. Russell Doolittle, chairman of the Chemistry Department at UC San Diego, is even more critical of the trends he sees. Asked if the direction of research is being altered by the flood of "new" money, he said, "I haven't got the faintest doubt. There are tremendous examples: my pristine *E. coli* friends suddenly became interested in blood-clotting when cloning came up—because they could begin cloning factors that there'd be a market for. It's extraordinary."

Doolittle, who "on principle" only consults for the government and nonprofit organizations, says, "I think that the university is the last vestige of a place where people ought to try to maintain some objectivity, where their choice of problems and how they go about solving them is not dictated by financial considerations."

In contrast, he has watched a steady erosion of proper goals: "An awful lot [of effort] is going to be put into making money, and that always means concentrating on the short-range. Nobody's taking the long-range view. I see an awful lot of gloom down the road after a quick fanfare."

Such thoughtless advance is bound to bring disaster, Doolittle feels. "Making money will go in two directions. Drugs will be first—most won't do any good or maybe will do harm." A study by the World Health Organization found that out of 3,000 licensed drugs investigated, "at most 250 did any good," he said. "Once a company has put a lot of money into developing a drug, it's very difficult to say, oh shucks, it didn't work." The tendency is to sell it as long as it doesn't hurt. "Or as long as it doesn't hurt *too* much." Meanwhile, says Doolittle, "The phenomenon of cancer is not understood. But it could be."

The second direction the money-driven research will take is agriculture, Doolittle said, and there "I've never seen any of these things *not* backfire, just like DDT." Noting that he believes the "nightmare" stories of the risks of recombinant DNA are overblown, he said: "No, the real hazard is that we're going to create a dustbowl doing something that *works*."

Friends tell Doolittle that federal money also has "strings attached," requiring concentration on specific problems. But "the main difference is that the NIH, NSF, and other agencies had not been working on profitability. Their vested interests are much more genuine than playing for bucks." Even so, he said, the government never has supported enough basic research. "There's a lot given to the medical world that I feel is ill-spent. Medical research is the flimsiest kind . . . high school research on a national level."

Instead of rushing after money to replace federal grants, he feels, researchers ought to learn to live with less. "It might make people do a little more thinking. The English have done great work on very little money. They sit around and talk more and drink more tea. But they've done very well."

Tom Poulos, a researcher at UC San Diego, also noted that while much corporate money technically does not require specific projects, "the tendency is to want to do something that the people who gave you the money would find useful." Further, the sudden prospects of great wealth have brought out the worst in some scientists, Poulos said.

"There are some real monsters around. I can't say whether they were nice guys who changed when fabulous wealth came their way, or whether they were always like that but it never showed. But I guess they're like the robber barons. Twenty or thirty years from now we'll say, 'I knew him. He created the billion-dollar ABC company. But he wasn't the kind of guy you'd want to know.' "

Poulos, a protein chemist, believes communication is shutting down. "People don't talk anymore. I can't stand that. It even gets to me, and eventually it gets to everybody. I've been working on this [protein chemistry] for a long time. Suddenly the work I'm doing might become commercially valuable. That'd be nice. I don't want someone to come along and pull the rug out from under me. So we all become a little different."

Even Stanford's Kennedy, while urging progress in the new relations, points to unwanted side effects. Referring to competition in science, he said in congressional testimony, "I cannot believe that the new terms of trade will not make it much, much worse. Whatever temptations to secrecy and competitiveness are now put in our way by human frailty can, it seems to me, only be amplified by adding the joint temptations of power and profit."

Stephen Atkinson sees a great deal of the research effort at Harvard as executive secretary of the Committee on Patents, and his fear is that few university administrators have any idea of the magnitude of change coming to their campuses. "This and other universities are asleep on their feet about how this is going to affect them," he said. "A lot of positive things are going to happen, yes, but my point is that they aren't aware of the change coming at all."

Evidence of that is the belief that the genetic engineering upheaval is not different from commercialization in the other sciences, Atkinson said. "The Chemistry Department has been consulting for fifteen years. Fine, that means ten or fifteen days a year for very distinct, discreet time periods, and they get paid well. But look at the difference: all these [biology] guys are getting *stock*. The degree of involvement has never been so intimate. You've got a much more intense involvement of these professors in companies, and that's going to make a substantial difference."

Referring to a Harvard professor known to be secretive and aloof, Atkinson said, "The problem isn't that he isn't talking to anybody anymore. The problem is that there are twenty people in his lab have made agreements not to talk to anybody. How does that affect students? What if they're onto good ideas? Before, if you stole a student's idea, that was plagiarism. Now they grab him and form a company around him."

A NEW DEAL?

There is another view of the sudden demand in genetic engineering: some say it simply represents a welcome shift from a talent-buyer's to a talent-seller's market. Many scientists—including those concerned about the new corporate influences—point out that for years Harvard and other top institutions had a monopoly on the nation's best young scientists. Harvard, for example, retained them for up to seven years, then turned them loose without tenure. The system worked in its time, because with Harvard credentials, those turned loose could almost always find posts at other good universities.

Now, many say, most of the best scientists still head for universities, some of the best look elsewhere, and other job-seekers still have a choice of good jobs. John Abelson, UC San Diego professor and a principal in the Agouron Institute, an rDNA firm, said the commercial opportunities came at a welcome moment. After the tremendous increases in enrollment of the 1960s and 1970s, university growth has dropped off sharply in the 1980s. In UC San Diego's molecular biology department, he said, an incredible 85 percent of the faculty are full professors, most of them fairly young. "The spaces are full. We have a declining student population and a decline in money. But our *field* is going wild. We have to have some way to get young people into it."

Abelson has seen large numbers of bright young graduates suddenly head for DuPont and Monsanto. "They want to be assured of facilities for research, and they're getting good projects to work on, too." Like Baltimore and Gilbert, Abelson believes the small rDNA companies are the place where practical, soon-to-be profitable problems should be worked on, the lab the place for more fundamental research. But he does not see a shift in the lab work. He agrees that, by and large, the top scientists have not left academe to join companies. He offers this reason:

"To leave research now would be the largest of mistakes. If you're not training graduate students, you are not at the forefront of the field, you've lost the fountainhead that will be the source of *future* industrial importance. Grad students talk to their peers, they are excited and imaginative. They carry their professors kicking and screaming into the next generation."

Gilbert, while still at Harvard, saw a split coming in biology that may make many of these concerns passé. "I still do very basic research in my lab at Harvard, but part of the *field* is moving away from that basic level, and that's the work the company does. Molecular biology on the whole I view as splitting. An applied field, including the whole of medicine, will go into the companies. The part of molecular biology stressed in universities is going to be much more about the underlying controls and development of life. The *structure* underlying evolution is now getting more and more discussion—how genetic structure influences evolution."

Moreover, Gilbert believes that attempts to stereotype researchers for pure or applied science are misguided. "One of the reasons you do basic research is that you are self-motivated to find long-range solutions to the problems of society. Now you suddenly come to the point where you can say, well, it doesn't *have* to be long-range, I could actually work directly on something that has a direct application to society. You may, in fact, wish to do that, and it would be part of that same overall view of the world. Is it more worthwhile for me to spend my time applying new techniques to problems that will be useful now, or to work on basic research that will have an impact in twenty years? There's not an automatically correct way of making that judgment."

For Gilbert's part, long before his year's leave of absence from Harvard had ended, he said that although uncertain what he would do at Harvard "I intend to remain chairman of Biogen."

At Cetus, Shing Chang suggests the form biotechnology might

take as the field splits, relating it to engineering, his first academic interest. "I'm partly hoping and partly predicting that an interchange will occur, there will be a turnover in both universities and industry, there will be a high demand for people skilled in basic or applied research. Therefore, if a university wants to train its graduate students for *both*, it should have teachers trained in industry as well as basic research. There doesn't have to be a formal exchange, it may happen as it does in engineering: people quit high-paying industry jobs to teach for a period."

But regardless of method, Chang is concerned that some means be found to support the most basic research. He publishes his findings, as does Genentech, he noted, but by and large scientists in the industry do not. Also, "We do as much basic research here as we can afford to," but most still must be done in universities.

Chang summarizes best the concerns of those trying to see the future of genetic engineering. If universities have had to come to grips with the demands of the marketplace, the companies now must see that their futures are being worked out in pure-research labs, where basic knowledge pursued without consideration of dollar value led to this industrial revolution that will have fabulous economic value.

Dangerous Acquaintances

The "worst case" scenario for recombinant DNA would play itself out as a true nightmare:

A research scientist inserts the genes that produce the deadly botulin toxin into a colony of *E. coli*. A few bacteria accept the gene, so they now produce the poison of botulism. Most of them, for one reason or another, are enfeebled by the new gene. One remains virulent and healthy, multiplying in its usual fashion. There is an accident—a dropped flask? an accidental brushing of the colony with a coat sleeve?—and some of the deadly *E. coli* infect the researcher, reproducing in his intestine, as *E. coli* always do, this time killing him with botulism. Unfortunately, the deadly *E. coli* are very good survivors; they spread through the city, over the countryside, without natural or artificial enemy or antidote, from person to person, in an epidemic unmatched in history.

Impossible? Most scientists now believe so—or rather, since nothing is absolutely impossible, that it is of such very low probability that in point of fact it will not happen, for reasons we'll discuss. But it was scientists themselves who first became concerned that such a devastation *might* be possible, who came together in a series of conferences to assess potential hazards in recombinant DNA investigation. The best known and most written about of these were two conferences at the Asilomar Center in California.

The leading molecular biologists in the world convened there, drawn together not so much in fear that the worst would happen as by uncertainty over what risk levels and odds of occurrence ought to be assigned to various experiments. One's fears are not based purely on likelihood of an occurrence, but equally on the gravity of the results if it *should* occur. Thus, even the early certainty that there was little chance of a recombinant DNA plague was not reassuring, since the perceived "slim chance" events would be so horrible. The scientists agreed to observe a moratorium on rDNA experiments until the National Institutes of Health—the funding arm for most of their experiments—could set up guidelines. The most hazardous experiments could be prohibited, those adjudged safe could be conducted with minimum regulation—a simple exhortation to use "good laboratory practices"—and various levels of care could be required for those in between.

The guidelines that followed have been in effect since 1976, and although the controversy in the scientific community subsequently died down, the debate is one that reasserts itself periodically. The NIH guidelines require laboratories to operate at "containment levels" coded from P1, for the least risky experiments, to P4, for those potentially most hazardous.

But in the summer of 1980, these guidelines were relaxed considerably. Dr. William Gartland, executive secretary to the Recombinant Advisory Committee (RAC) in the NIH since its formation, explained that the categories have not been altered, but vastly more experiments now may be conducted at lower containment levels. Even more controversial is a proposal by David Baltimore, a member of the RAC, to make the guidelines completely voluntary.

Most experiments with *E. coli* have been deemed safe, and are thus exempted from these guidelines. Gartland says that means that "the lion's share of DNA work is already technically exempt from the guidelines."

Why? First, the worst case scenario makes several assumptions, notably that an *E. coli* strain, weakened by having to make foreign proteins of any kind, would be able to compete "in the wild" with strains that have evolved over millennia to become adapted to their environments. Further, it assumes that a single colony of such a strain would not merely find an ecological niche—that is, survive as a deadly organism—but would succeed *better than* natural competitors. Since the new *E. coli* would kill its host and thus, ultimately, itself, that does not appear logical.

But this particular point should not cloud the issue: many other scientists who have followed the debate since the beginning do not believe enough attention has been paid to a variety of risks that go beyond the simple if terrifying accident scenario, nor do they feel scientists have made a serious enough attempt to answer long-term questions about the effects of the technology. The debate has divided the advisory committee itself.

Ray Thornton,[1] president of Arkansas State University, told other members at a 1980 meeting, "That argument that guidelines for the conduct of publicly funded research are not needed because nothing more dangerous than now exists may be developed by this technology is beguiling, but false on two counts:

"First, human experience has shown that any tool powerful enough to produce good results of sufficient importance to shake Wall Street and offer hope of treating diabetes is also powerful enough, wrongly used, to produce bad results of equal consequence.

"Second, even accepting the assumption, with which I disagree, that nothing other than beneficial results may stem from this research, it does not follow that the public thereby loses its interest in having suitable guidelines for the expenditure of public funds. . . ."

The problem has been caused partly by a failure to distinguish the various *kinds* of risks in genetic engineering, and many feel that the focus on laboratory hazards has masked more real dangers. We can think of four categories of risks that ought to be associated with biotechnology.

- *Accidents.* The guidelines were set up to deal with laboratory hazards, to prevent experiments from being carried out without proper safeguards, which might result in the escape and survival of a dangerous mutant strain. Most scientists now believe that early fears on this score have proved unfounded.
- *Long-term damage.* Little debate has been given to the possibility that genetic engineering will have serious, adverse consequences to the environment, leaving a

legacy as full of problems and health hazards as the chemical industry has.

- *"Mega" environmental damage.* Will genetic engineering in humans, which we hope might eliminate dreaded genetic diseases, ultimately monsterize the race? Or will we find that one group of people might try to dominate the determination of which genetic traits are "best?"
- *Deliberate damage.* There are fears among some scientists that genetic engineering could add a new and terrible dimension to biological warfare, that the gravest threat of damage is from those who would twist powerful new techniques to destructive ends.

Clearly, the deliberate use of the powerful new recombinant DNA techniques to create deadly plagues or to otherwise incapacitate or mutilate entire populations is the most terrible potential risk for the future of genetic engineering. It is also the one neither we nor anyone else can assess, because whether such research is being carried out by the United States or other nations is kept from public knowledge. As we assess the other risks of genetic engineering, it should be kept in mind that we draw a blank in trying to foresee what potentially seems the very worst to us.

GREAT ESCAPES

The worst case scenario painted earlier should be as familiar as it is frightening, for it has been drawn and redrawn since the nineteenth century, when the achievements of science became the stuff of fiction and prophecy. When Mary Shelley wrote *Frankenstein*, scientists *were* making muscles contract with electric shock, and they *were* watching corpses' hands open and close. But the Frankenstein monster escaped from the lab and into our culture only in fiction.

Similarly, long before Michael Crichton wrote *The Andromeda Strain*, science fiction writers had harvested stories based on new diseases and monster mutants, although until now their favorite prime mover for the disaster has been radiation-caused mutation.

Nevertheless, the reality of biological research is in some ways more surprising than fiction.

There are *no* regulations, federal or otherwise, governing experiments with even the deadliest microorganism, such as those that cause Lassa fever, bubonic plague, or typhus. The NIH guidelines, such as they are, apply only to experiments with recombinant DNA, not the natural kinds of bacteria that cause plagues. The NIH's Gartland pointed out that although the government produces a manual of suggested measures for dealing with various pathogens, there are no laws on the subject. There are similarly no federal laws dealing with carcinogens. Containers of hazardous substances shipped interstate must be so labeled. "But you can go home and work in your kitchen on any kind of deadly toxin, and you're not breaking the law," Gartland said. "There is a certain irony that you have this whole regulatory scheme over a theoretical hazard, and no regulations over proven hazards."

Over the past century, laboratory workers *have* occasionally fallen victim to the pathogens they were studying and some have died from their infections, but not once, despite the total lack of regulation, has a laboratory infection spread to the public.

Little news value has been attributed to such laboratory infections, yet they are grim enough. In 1976, Robert M. Pike[2] of the University of Texas Southwestern Medical School, Dallas, published a survey entitled "Laboratory-Associated Infections; Summary and Analysis of 3,921 Cases." Pike gathered information from published reports and from questionnaires sent to researchers as early as 1950. Of his nearly 4,000 cases beginning in 1915, 2,465 occurred in the United States, and 164 of the infections were fatal. A later update by Pike added 158 cases, six of them fatal.

The diseases were as exotic as any imaginable: "Q fever," so named because it raised so many questions among researchers, a disease that is seldom fatal but that can lead to pneumonia; Venezuelan equine encephalitis, tularemia, brucellosis, and psittacosis; hepatitis and typhus; Marburg fever, which mysteriously felled laboratory workers in Marburg, Germany, in 1967, killing seven of them. There is "B virus," which Pike said, "has caused more deaths among laboratory workers than any other virus and it is second to the typhoid bacillus in fatalities due to all agents." Lassa fever, arising from a virus endemic to West Africa and causing inflammation of the organs, "appears to be one of the most hazardous viruses in the laboratory, proving fatal in two out of three laboratory-acquired infections."

Some have occurred despite precautions, others through carelessness. H. T. Ricketts, whose work in nutritional ailments lent his name to one of them, died in Mexico while studying typhus after lice escaped from an envelope in his pocket on the way to the laboratory.

In 1970, a woman became infected during an autopsy in a Lassa fever victim in Nigeria. "When the newly discovered virus was brought to the U.S. for study, a female laboratory technician in New Haven, Connecticut, working with the virus was fatally infected. Another worker in the same laboratory suffered a severe infection but recovered," Pike said.

Q fever killed one researcher during a massive outbreak at the NIH's own facility in which 153 workers were stricken, and eight years later a second outbreak infected forty-seven people. In that last episode, six employees of a commercial laundry patronized by one of the workers became ill, and they remain the only "outsiders" known to have fallen victim to a laboratory-initiated disease outside the immediate families of a few researchers. Such an accident would be highly unlikely today because of stricter observance of rules governing lab clothing. Pike pointed out that nearly all the infections occurred before the use of ventilated lab hoods and before sterilization of lab clothing and mandatory showering, all now routine.

At about the same time, Dr. A. G. Wedum[3] of the NIH did a similar survey that was to have major implications for rDNA research; he studied the history of infection at Fort Detrick, Maryland, the army's biological warfare center. Also published in 1976, the study evaluated the chances of an rDNA accident if Fort Detrick's maximum-safety equipment were in use. That equipment would become the basis for defining the P4 maximum containment level in the NIH guidelines.

In various P3 and P4 labs at Fort Detrick between 1959 and 1969, Wedum found one or no infections, and those that did occur were outside the labs' protective devices. One infected worker had been smoking while working at an open benchtop, another brushed his own infected clothing after changing, and still another orally pipetted an antigen and accidentally swallowed some. None died. (While the study indicates that the protective devices did not fail, it does not explain why these workers were not using them.)

Physical protections in the higher containment levels include protective clothing such as gloves, ventilated hoods that draw off airborne pathogens, fully enclosed glove boxes that prevent any contact with the pathogens whatsoever, and negative air pressure in the lab—that is, a slight vacuum—to prevent currents from wafting out. Coupled with this, rDNA researchers now are using enfeebled strains of *E. coli*, such as one named K_{12}, deliberately developed so that it cannot survive outside the laboratory.

Further, many scientists believe that it would be extremely difficult for them to *deliberately* create a recombinant organism that would stand a chance of competing in the wild, and that even the deadliest of recombinants designed with protection in mind would die outside its petri-dish culture. Given the precautions, they rule out the kind of widespread calamity once envisioned. Ardyth Myers, community relations director for Genex Corporation, noted that one scientist predicted, "We'll someday need our P3 lab— to protect our strains from being killed off by a hostile outside world that might get in here, rather than the other way around."

Bacteria mutate all the time naturally, but so few mutants survive that in nature, mutation is generally synonymous with death. On the rarest of occasions, a natural mutant is suitably enough adapted to its environment that it can win its own niche among all the strains already competing there. Artificial recombinants would stand still less chance of surviving.

But even among those who feel the major threat is not from laboratory accident, there is dissatisfaction with the steady erosion of the guidelines' force.

Sheldon Krimsky, a member of the RAC until recently, agreed that the likelihood of a disastrous accident is small and that "history shows that it is not the laboratory which produces the greatest damage." He noted that in terms of biohazards the governmental "institutions developed to look into them were formidable, fair, and reasonable, even though I didn't agree with all the decisions made within those institutions."

Krimsky, a nonscience member of the RAC, is a professor of urban and environmental policy at Boston's Tufts University. A frequent critic of efforts to weaken regulations, Krimsky was not reappointed to the committee when his term expired. His major complaint is that, while everyone talks of the unlikelihood of disaster, "I see no efforts made to see if the worst-case scenario would work—putting the genes for botulin into *E. coli*, and really trying to *see* if under extraordinarily controlled circumstance, you *could* get it to survive and multiply. They say they couldn't, but when they begin to do experiments like that and do them honestly, I'll begin to be convinced that this stuff is not less hazardous than botulism all by itself. But those experiments have not been done. They're fearful that if they do them, they'll find out too much."

By contrast, Krimsky says, "if anything, the Army is likely to do those dangerous things."

That uneasiness extends to others who generally applaud the work of genetic engineering and its promise for the future. Bruce Hilton, director of the National Center for Bioethics in Richmond, California, said that despite reassurances of safety, "it still gnaws at the back of my mind: have those fears really been resolved, or is this a drawing together of members of the science community, fearing that they may confront another Galileo- or Copernicus-situation unless they pull together." His uneasiness is heightened by sensing "terrific pressure from the pharmaceutical companies and labs to get on with it and not be tied down with restrictive research rules." In any event, he says, "They seem to relax the rules more every six months."

Gartland, an executor rather than formulator of policy, is cautious in assessing the changes and proposals. But he notes that making compliance voluntary "in effect marks a dismantling of the whole system, and that has some people worried. Because if you dismantle it and something happens, then we are back to ground zero again: 1974. You've got no national committee and no national forum for anything, and that's a very dangerous thing."

Perhaps most importantly, the guidelines now apply only to work done under federal contract. As Gartland puts it, "All we can do is 'follow the federal dollar,' to say that following the guidelines is a condition for your getting this money." As more and more recombinant DNA work goes into uncontrolled industry, the limitations of such regulations make less sense. Industries have agreed to follow the guidelines, but that already is voluntary, except in a few cities that have ordinances governing genetic engineering. Gartland pointed out in late 1981 that a multitude of congressional efforts to pass national legislation regulating recombinant DNA work has gotten nowhere. "It was decided that no federal agency had complete jurisdiction—the Occupational Safety and Health Administration comes closest, because it can require that you provide a safe workplace." But even OSHA has no specific charge with regard to laboratory safety. "Therefore, national legislation was needed. But no law has ever been passed. It remains in limbo to this day."

The lack of monitoring of industrial recombinant DNA work is an important issue because many believe that the concentration on possible accidents, while well-intentioned, is masking a far greater hazard: the one we haven't foreseen, the obvious one we never seem able to see in the world around us.

TIME BOMB

UC San Diego's Russell Doolittle argues that, just as organic chemistry and other branches of science rushed too quickly to the marketplace, industrial genetic engineering will be marked by a brief period of bright hope followed by environmental ruin. We won't likely set loose a deadly new bug. "We'll create a dustbowl where a farm state used to be. It won't happen because something crept out of the lab. It'll happen because in our rush, we didn't think of all the consequences of doing something that looked good at the time—something which in the short run will *be* good."

For Doolittle, proper precedents to consider would be the pollution of the atmosphere by automobiles, or the terrible erosion caused by slash-and-burn techniques in creating new farmland. "Who watched that first internal combustion engine roar and thought, 'I wonder if that'll clog the atmosphere with junk, kill plants, foul rivers?' Who could have? I'm afraid that in this case just such mundane failure to look ahead will cause the real disasters."

Sheldon Krimsky also believes that, far from being safer, industrial scale-up is precisely the action that will usher in new hazards through "the purposeful dissemination of materials in the environment with inadvertant consequences."

Bruce Hilton asks, if plants that can fix their own nitrogen come to dominate even a branch of agriculture, could they also change the nitrogen balance of the air? If so, what would the consequences be? If that seems farfetched, Hilton notes, who would have believed aerosol-spray cans could affect the ozone layer?

The suggestion made repeatedly by those concerned is not that investigation be halted, but that the long-term implications of genetic engineering *applications* be studied. Krimsky is especially critical of the government in this respect. Of such study, he says, "The one institution that we have in this country designed to do that, the Office of Technology Assessment, has completely relinquished any of its capacity to do so."

He said the OTA's "fundamental interest has been: Can we compete with other countries? and, How can Congress stimulate the development of biotechnology?" Krimsky adds, "That's perfectly legitimate, there's nothing wrong with that, but it has to be balanced with some assessment of how these things are going to alter society. Do we want to undertake a system for producing biological pesticides when we had such an outrageous situation with organic chemicals?"

Organic chemistry, moreover, produced few hazards in the *laboratory*—certainly none that could not be handled. It was as an industry that it left us with disaster along the Love Canal in New York and contaminated municipal wells in the Southwest, not to mention providing those once universally favored aerosols.

Further, with sharp cuts in federal funding, many universities are turning to other revenue sources to back their investigations. Since the NIH guidelines apply only to NIH grants and are otherwise voluntary, fewer and fewer experiments may come under these minimal controls, let alone regulations that would further restrict industrial applications. Gartland noted that the only rule that could now apply prohibits production of recombinant DNA in batches larger than 10 liters when the DNA is not "well-characterized," that is, when its protein product is not known. "But it's hard to imagine you would want to scale up production if you didn't know what was going to come at the other end," Gartland notes.

There is even economic dislocation that raises some of the same ethical questions as do potential hazards. William P. O'Neill, a vice president of DNAX, a Palo Alto gene-splicing firm, points out that human blood serum is now used to make a multitude of products, including the vital clotting factors that keep hemophiliacs alive. The cost of the factors is underwritten by other, widely sold products. "What happens if next year we genetically engineer enough of these products that there's no market for human serum?" O'Neill asks. "Do we start charging hemophiliacs $1 million for a treatment?" O'Neill said recombinant DNA also might provide those clotting factors, and at an even lower cost than they are now available, but he fears few people have considered the need to coordinate those developments. O'Neill has spelled out these concerns in an appendix volume he wrote to the OTA study.[4]

The pressure to extend the NIH guidelines beyond the federal grant system and to monitor industrial development has taken many forms, but so far its only fruition has been through several local ordinances. The first, and the one that set off the most controversy, was that in Cambridge, Massachusetts.

GUARDED WELCOME

Cambridge became the focus of controversy over rDNA experiments almost as soon as the NIH guidelines were published, and with good reason. As home of both Harvard and M.I.T., the city fathers correctly assumed, Cambridge would become home to many rDNA firms that would not have to obey the federal guidelines. The battle on the city council was characterized by as many wild, heated charges as logical arguments, with then-Mayor Alfred Velucci forecasting that the city would be plagued by monsters and epidemics. For all the fury of the public hearings, however, Councilman David Wylie recalls that only a handful of people were present on April 27, 1981, when the council passed the nation's first ordinance[5] governing recombinant DNA, and it is a marvel of simplicity.

Basically, the ordinance requires anyone doing genetic engineering research to agree to follow the NIH guidelines and to get approval from an advisory committee of the council before doing recombinant DNA work. Both laymen and scientists make up that committee. In 1981, the ordinance was updated to cover the larger scale rDNA work expected from the cluster of new firms in the city; they must now get licenses and follow newer NIH guide-lines governing recombinant experiments in larger quantities.

Macy Koehler, who became Harvard's biology safety officer when the post was created in 1977 and served until 1979, praised the ordinance both because it will protect the public from acci-dents and because it will protect researchers from misunderstand-ings about their work. As safety officer for those first years, Koehler made sure that not only the NIH rules but university biosafety rules were followed throughout the Harvard Biological Laboratories.

"I believe in public involvement in scientific decision making," the biologist said just after the ordinance went into effect in the summer of 1981. "We must make an effort to communicate to nonscientists, to tell them what we're doing and why. But non-scientists must make an effort, too. Even when put in the very simplest language possible, the subjects [of genetic engineering] are very difficult."

Boston soon followed Cambridge with an ordinance, and sev-eral more cities have adopted them since—but they remain a handful nationwide. To Councilman Wylie, that is a great weak-ness in the safety net. "There is certainly no reason to be hopeful of industry on the whole [being careful] because of public apathy," he said. "For every well-run company, I fully expect to see one the opposite. Look at arms manufacturers: they develop weapons for national defense and sell them all over the world as quickly as possible and for as much profit as possible with no thought to national security. Why expect things to be different if the public is going to remain indifferent to what its government is doing?"

Wylie is a strong supporter of genetic engineering as a new in-dustry for Cambridge and of the ordinance as proper regulation of it. "We need new industry," he said. "Cambridge lost a lot to urban renewal," including the city's largest employer, Simplex Wire and Cable Co. "Nobody was really saying we should keep the new industry out. I felt it was healthy for Cambridge to take this kind of care."

On the other hand, Wylie feels that the federal government has been extremely lax in protecting the public health against all en-vironmental threats. "The only means of continuing democracy lies in citizen participation," he said. "Here we have a case of one city putting the federal government to shame. The government has become weak, insignificant, immobilized by special-interest groups."

Even Krimsky, who is generally not as critical of the NIH as Wylie is, feels it is a grave error to have industry-involved scientists sitting on the Recombinant Advisory Committee. Citing a Boston *Globe* report that David Baltimore could sell his holdings in Collaborative Genetics for more than $4 million, Krimsky said, "To think that he's spearheading the deregulation of the guidelines! I don't care how honest and honorable a person is, there's just the appearance of such a great conflict of interest that in any other time and industry I don't think anybody would buy it. Imagine someone with a large block of General Electric stock sitting on the Nuclear Regulatory Commission. If the public found out, they'd be outraged."

Krimsky says a problem has always existed on the RAC to some degree because it had to have research scientists on board. "But I would argue that there is a much deeper conflict of interest when a person can make millions of dollars on decisions that they're involved in" than when they have only an intellectual or academic interest. Krimsky's term on the committee expired in June 1981. He was renominated by NIH Director Donald S. Frederickson, but that was vetoed by someone in the former Department of Health, Education and Welfare. Krimsky says most committee members leave when their terms expire and that it is the administration's prerogative to name members to the committee, but that it is *not* usual for nominees to be turned down. "I was probably the most cautious, and I probably spoke out consistently on problems of regulation more than anyone else," he said.

ENGINEERING EVOLUTION

To many scientists and religious leaders, even the fears of environmental damage are dwarfed by the major ethical questions raised by genetic engineering, by such possibilities as our altering the human race significantly to create new kinds of humans. To repeat Shakespeare's phrase with Huxley's construction, "O brave new world that has such people in it." Some of the fears voiced have been hysterical. The U.S. Constitution, as we said earlier, cannot be repealed by scientists. But real pressures can be put on all of us to change that Constitution, and many fear we are by no means prepared to deal with the challenges we face.

Genetic counseling is already reality, ushered in by the ability to detect key genetic factors through amniocentesis and the Supreme Court ruling permitting abortions. Bruce Hilton is such a genetic counselor, a Methodist minister who spent two years at the Hastings Institute during which he was senior editor for the book *Ethical Issues in Human Genetics*. He now publishes a newsletter on bioethics as well as directing the center. It was Hilton who referred to possible "mega-environmental impacts" of genetic engineering as "short-circuiting evolution."

"What happens when you jump several million years from one organism to the next? Maybe nothing, but nobody in science seems to be addressing questions like these," he says. Hilton says he rarely mentions that he is a minister because "scientists are suspicious of religion's old role" in suppressing inquiry, and he is very much in favor of the progress promised in biology. "I'm in favor of amniocentesis and a woman's right to an abortion," he says. But he sees dangers lurking, some of which already have revealed themselves.

For example, some states have tried to require black couples applying for marriage licenses to be screened for the sickle-cell anemia gene. The couples may marry whether they carry the gene or not, but Hilton sees such forced screening as a possible prelude to something far deadlier—either legal bars to such marriages or forced abortion. He pointed out that 8,000 babies are born in the United States every year with Down's syndrome (mongolism), and it costs about $250,000 to institutionalize each of them over the course of their lives—that is, about $2 billion a year figured over an indefinite period. Half these babies are born to an identifiable group—women over forty years old. Hilton is concerned that eventually there could be pressures to force women identified as carrying Down's syndrome babies to have abortions, to save $1 billion a year of those costs.

"I certainly would not be in favor of any such idea. Society can bear the cost more easily than it can bear such a striking seizure of the freedom to procreate," Hilton said. "But these pressures would not be new." Further, he said, "There are some serious, thoughtful people on the other side."

Charles Frankel wrote in *Commentary* in 1974 that a deliberate intention to create a "new man" unites all revolutions of the right and left. "These are the accents with which Sir Francis Crick speaks when he states his belief that no newborn infant should be declared human until it has passed certain tests regarding its genetic endowment, and that if it fails these tests it forfeits the right to live."[6]

More pointedly Frankel says, "The partisans of large-scale eugenics planning, the Nazis aside, have usually been people of notable humanitarian sentiments. They seem not to hear themselves. It is that other music they hear, the music that says there shall be nothing random in the world, nothing independent, nothing moved by its own vitality, nothing out of keeping with some Idea: even our children must be not our progeny but our creations."

Such arguments are, indeed, all part of a larger view fought over for much of this century concerning "eugenics," and those favoring strict procreation control formed an extremely powerful group that predated Adolph Hitler, but one which is presumed to have given him many of his notions on race and racial purity. Eugenicists have included presidents of major universities, leading liberal intellectuals, industrialists, and political leaders, and they were extremely influential through the 1920s. At the heart of their thinking, Hilton noted, is a belief in human perfection. He added, "It is part of western lay theology that it is possible to cure death and that it is possible to make a perfect human. Both are wrong, but I'm convinced these notions exist in our subconscious."

Hilton's point: The major teachings of eugenicists may have fallen out of favor, but they remain near the surface, and there is reason to be concerned that genetic engineering and other forms of biotechnology will bring them to the fore again unless there is adequate discussion of the issues. Certainly, genetic engineering might detect the crippling diseases Watson referred to at such an early stage that all but those who oppose abortion on purely religious grounds would come to strongly favor it.

Hilton said that R. G. Edwards, a British pioneer in developing the technology for the first "test-tube baby," told him that human embryos can now be "typed" for genetic illnesses and sex when they have split to only eight cells. Further, if it becomes possible to alter the genetic *makeup* of embryos in some early stage of development, for the first time in history genetic illnesses and many birth defects would be curable.

It might seem hard to imagine a test that would detect pregnancy at such an early, hours-old stage as an eight-celled embryo. But if eventually one were available, it would be a boon to women at high risk of bearing defective children.

The coming of the Gene Age will not find us wholly without legal guidance on these major issues. Experimentation with human subjects, for example, has long been strictly regulated by the government, and FDA requirements for bringing new drugs to market are estimated by one man familiar with the business to take an average of eight years to fulfill. But neither are the questions idle or premature. We appear to be on the verge of cutting the umbilical cord to nature, freeing ourselves from muscle power and industrial power, perhaps even from the ties that bind us to the natural and slow process of evolution.

Ray Thornton, in his statement to the RAC, said people are most comfortable with science "which explains how things work, which promises health, physical well-being, and material progress." But now and then, he reminded, science bursts its bounds and leaps into areas that challenge our concepts of life and the institutions we have built in service of our lives. "When Galileo offered the theory that the Earth revolves around the sun, it was bad enough to his contemporaries that [in their opinion] he committed scientific error. It was worse that he committed heresy as well."

The questions of genetic engineering that most chill our souls, too, involve not error or chance, but the violation of our concepts of life itself—heresies to believers and atheists alike.

Genetic Engineering: The Science III

The genetic code. We live and die by it, procreate, grow, mutate, change, or remain the same according to its dictates. This is no code of conduct, but something closer to a linguistic code: rules of expression and rules for literally translating the four-letter language encoded in the bases of the DNA chain into a quite different language of amino acid sequences that compose proteins. We and the rest of the living world are made with significant amounts of protein and virtually all our bodily chemistry is carried out by enzymes, which are just a particular kind of protein.

We're going to look more deeply into the process by which DNA *instructions* lead to living *things*, a process still being unraveled by molecular biologists. But first let's take another look at the highlights of that DNA language system.

A DNA molecule is composed of two long chains twined around one another, joined together by complementary bases like steps in a staircase. The order of the bases determines a set of instructions just as the order of letters in English does. The region containing the instructions for a single protein is called a gene. Messenger RNA (mRNA), a molecule related to DNA, is composed of similar bases, and such a molecule is built to copy a particular stretch of DNA using a cellular copying machine called RNA polymerase. The messenger then travels to the ribosomes, the cell's automated workbenches, which "read" the orders from mRNA's base sequence and translate them into a particular sequence of amino acids, the building blocks of proteins. As we trace the procedure in detail, we'll follow all the way through the translation process using the genetic code.

Through the schematic microscope, the view in Figure 7 shows a DNA sequence of a hypothetical *E. coli* gene. In front of the gene region is the promoter, or control sequence. Next to the promoter, pictured below the DNA sequence, is the mRNA copy that has been formed, and below that is the sequence of amino acids that will be called for by that messenger RNA. Notice that the two DNA strands have different names—template and coding —and that makes sense given the principle of complementarity. The coding strand has the *same* sequence as the messenger RNA, except for the substitution of the base uracil for thymine in mRNA. The other strand is called the template because the RNA polymerase copying machine actually works by *reading* the template strand, then assembling its complements. That forms a messenger RNA identical to the coding strand of DNA.

The figure shows several special sequences that must be read to produce the mRNA copy: the promoters and the start and stop sequences. The promoters are sites recognized by RNA polymerase. The sequences at these sites bind to the RNA polymerase, allowing it to "sit down" on the DNA to begin copying (transcribing) mRNA farther downstream (to the right in the figure) on the molecule. Most promoter site sequences in *E. coli* are similar to those shown—similar, but not necessarily identical to one another, another interesting property. For example, this sentence isn't right: "The qik bro fx jumpeed ov thr lazi dig." But

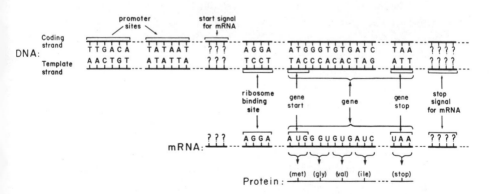

Figure 7. Reading DNA language.

despite the many typos, it contains enough information for you to recognize it. Similarly, many promoters are alike enough to be recognized by RNA polymerase—although here, no one of them is "correct English," while the others contain errors. All are near variations of one another. If you were handed a piece of DNA, in most cases you, too, would be able to spot the promoter. In *E. coli*, the promoters will contain sequences identical to or similar to those pictured here. For example, they all contain a version of the "Tata Box," that is, the sequence TATA—that's how investigators recognize promoters. Recognizing promoters is important because they indicate a gene nearby downstream, and you would be able to find it. Furthermore, you would even be able to say what protein would be made on the ribosomes from that gene, once you've learned to speak in genetic code.

The sequence of the bases in the promoter region determines how strongly RNA polymerase binds. Thus, if the promoters are strong ones, at any given time more RNA polymerase will be making messenger RNA. Genetic engineers are beginning to learn a lot about which promoter sequences are strong. In nature, cells often have strong promoters in front of genes for proteins they need in large quantity, and weak promoters in front of genes that are rarely expressed—although this is not the only method of controlling protein production.

The remaining two "signal sequences " are the "start" and "stop" signals. They tell the RNA polymerase to start copying just before the beginning of the gene as it moves along the DNA and to stop just after the end of the gene. In Figure 7, the start signal is shown by question marks because there are several variations of such signals. Often the sequence . . .CAT. . . is found near the start, but often not. We know little of the stop signals, but there are some stop regions rich in G and C bases. We do know that reasonably definite start-stop signals exist. Thus, the mRNA copy of the gene contains the base sequence coding for the gene plus some bases on either end of the gene.

Now let's study the mRNA copy in Figure 7 more closely. You will notice that the sequence . . .AGGA. . . is the ribosome binding site, and, with the uracil replacement of thymine, we can guess that the mRNA will bind at the ribosome's complementary sequence . . .UCCU. . . so that protein expression can begin.

Other sequences are associated at one time or another with ribosome binding, but AGGA is the most frequent.

The strength of the binding site is important, and genetic engineers are trying to isolate other strong binding sequences, because if the bond is weak, the mRNA can fall off the ribosome before the protein is started. Again, having located the promoter sequence, you could find the ribosome binding site easily by looking downstream for the AGGA sequence; you might miss it, though, if you saw only an AGG sequence or another simple variation that, like the promoter, is close enough to be readable.

Now we come to the most important sequences of the mRNA, those which code for the various amino acids that will be assembled into a protein: the gene sequences. Since mRNA is composed of a four-base alphabet and proteins are composed of various combinations of these twenty amino acids, we need to learn to translate DNA language into protein language. Here it comes.

A BIOBLITZ COURSE
IN GENETIC CODE

Given a sequence of mRNA bases that has been transcribed from the DNA gene, what amino acid will be ordered up? The answer turns out to be simple, using the mRNA-protein dictionary that follows. This dictionary is literally what molecular biologists mean by the genetic code. Learn to use the dictionary and you can read any gene for the protein it will express. First, in the genetic code, all "words" are composed of three base letters. That means that three mRNA—or DNA—language bases (also known as a codon) determine one amino acid. For example, anytime you see the sequence GGU on mRNA, one molecule of the amino acid glycine will be incorporated into the protein being assembled on the ribosome, like another bead on a string. If the mRNA sequence reads GUG, the amino acid ordered is valine.

The genetic code dictionary for the twenty amino acids is shown in its entirety in Table 2 (next page), and here are a few examples in reading it. Consider the mRNA triplet AUG. The table says if A is in the first position, move to the third row down under the

Table 2. The Genetic Code

First Position	Second Position				Third Position
	U	C	A	G	
U	Phe	Ser	Tyr	Cys	U
	Phe	Ser	Tyr	Cys	C
	Leu	Ser	Term	Term	A
	Leu	Ser	Term	Trp	G
C	Leu	Pro	His	Arg	U
	Leu	Pro	His	Arg	C
	Leu	Pro	Gln	Arg	A
	Leu	Pro	Gln	Arg	G
A	Ile	Thr	Asn	Ser	U
	Ile	Thr	Asn	Ser	C
	Ile	Thr	Lys	Arg	A
	Met	Thr	Lys	Arg	G
G	Val	Ala	Asp	Gly	U
	Val	Ala	Asp	Gly	C
	Val	Ala	Glu	Gly	A
	Val	Ala	Glu	Gly	G

label First Position of the table. The second position in our mRNA triplet is U, located at the first column in the table's Second Position listings. Now you should be looking at the listings:

Ile
Ile
Ile
Met

The third position of our triplet is G, so from the instructions at the right-hand side of the table, you can see that the amino acid coded for is methionine (Met). Work out a few more for familiarization and find these answers: UUU codes for Phe (phenylalanine), GUU for Val (valine), GUC also for Val, AGC for Ser (serine), and AGA for Arg (arginine). How come *both* the triplets GUU and GUC will yield valine? We'll explain that later, along with the headaches that type of duplication might cause genetic engineers.

All that remains in learning to read DNA language is to learn the stop and start signals encoded in mRNA for making protein. It turns out that AUG is almost always the start signal for the ribosome. That is, when the ribosomal machinery hits AUG, it picks up a methionine amino acid and begins the string of amino acid beads. Therefore, almost all *E. coli* proteins begin with a methionine. The machinery continues reading bases in sets of three until it reaches a stop signal, coded for by the three triplets UAA, UAG, or UGA—shown as "term" on the table, for terminate.

Here's how all this translates for genetic engineers. They want to transfer the gene for human interferon into *E. coli* so the bacterium will make vast quantities of the rare human protein. They have tentatively identified a cloned colony containing the interferon gene. To make sure the desired gene is there, they will determine the base sequence of the area of the DNA around the gene using the experimental technique of DNA sequencing, and will thus know the corresponding mRNA sequence. After locating the start signal, they can translate that mRNA sequence into an amino acid sequence to see whether it is the one for interferon. Moreover, by sequencing the DNA *up*stream from the gene region, they can be sure that the ribosome binding site and promoter are at the best distance from the gene to achieve maximum mRNA production and maximum ribosome binding strength.

The actual translation process at the ribosome is tortuously complicated, but luckily we can understand some of it fairly quickly, remembering the principle of complementarity in binding. The mRNA is moved along the ribosome so amino acids can be tacked onto the growing chain, but the *actual* translation from RNA to protein language is carried out not by the ribosome but by other adapting molecules, another species of that varied RNA molecule called transfer RNA—tRNA. It is the tRNA molecules that actually carry the individual amino acids to the ribosome. Thus, the primary function of the ribosome workbench is to bind and orient both mRNA and tRNA so that the protein can be formed.

There are many varieties of tRNA already "made up," unlike the mRNA, which is assembled as needed. Each variety of tRNA recognizes two things: one amino acid and one mRNA base triplet. When found on the scene of protein formation (Fig. 8a), the clover-shaped tRNA already has the appropriate amino acid bound to it, just as with a little yard engine that is specific to one and only one kind of freight car. Figure 8a shows two of the many types of tRNA—one specific for phenylalanine and one to alanine. The "real-life" tRNA chain folds back on itself in a three-dimensional structure not unlike the clover leaf shown. While this folded tRNA structure is complicated, we need only look at two features. First, the anticodon region, so named because it is complementary to the codon region of mRNA. In phenylalanine tRNA, that region consists of an AAA sequence, at the tip of the middle clover leaf (the tRNA molecule has about seventy-five to eighty other bases not shown). Naturally, that makes it complementary to an mRNA's UUU sequence, and that is where it will interact with the mRNA. Similarly, the CGG anticodon on the alanine tRNA complements the GCC codon on mRNA, and this tRNA carries alanine bound to itself. Check these codes out in the table.

In Figure 8b we are witnessing the assembly of a protein—a glob, which is what most proteins look like—with another amino acid, phenylalanine, being added on. First, the mRNA lines up on the ribosome so that the codon for phenylalanine (UUU) is exposed to the tRNA anticodon. Then, the tRNA lines up by binding to the codon through the principle of complementarity, so that the amino acid it is towing is in proper position to be hooked onto the growing amino acid chain by an enzyme. In Figure 8c the mRNA has moved down one notch (three bases) on the ribosome, so that its triplet GCC, which orders up alanine, can be read. The tRNA for alanine positions it in just the right place for tacking on. In this case, the alanine is tacked on to the phenylalanine amino acid.

This process is as marvelously precise as any industrial robot's reading and carrying out instructions from punch tape.

We mentioned the headaches caused genetic engineers by the fact that several messenger triplets can order up the same amino acid. The problem: Not all organisms have all the transfer RNAs. Here is what happens: say the mRNA from an animal cell sometimes uses the code CUU for leucine, and other times uses UUA for it. But the bacterium in which the genetic engineer wants to insert a gene from the animal cell *only* uses the triplet

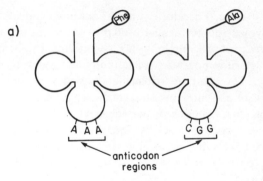

a)

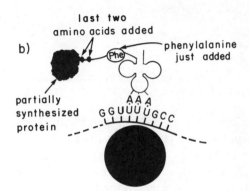

b)

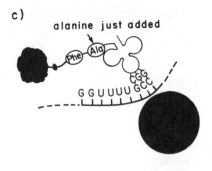

c)

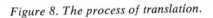

Figure 8. The process of translation.

UUA for leucine. Therefore, it might not have a leucine tRNA that will recognize the messenger's CUU, which means it can't synthesize any protein containing a CUU triplet. In effect, we're saying the bacterium doesn't have a complete, unabridged translation dictionary. The multiple triplets coding for one amino acid can be thought of as synonyms, and in this case our humble bacterium recognizes "car" but has no idea what "automobile" means. Nevertheless, once the lack is uncovered, it may also be possible to engineer into the bacterium the instructions for making the CUU-recognizing tRNA—that is, for making its dictionary more complete.

At this point, you know enough genetic engineering and its basic science to understand much of what is done in the field. But if you have a deeper interest, read on.

ADVANCED GENETIC ENGINEERING:
cDNA Cloning, Enzymes, Secondary Metabolites, and Scale-up

Higher organisms, such as animals and plants, have a more complex gene structure than bacteria, and that means different procedures have to be used to clone these genes into bacteria. The only technique now available to get the more complex genes is copy-DNA (cDNA) cloning. Shotgunning doesn't work because of a strange, unexplained trait of many of the genes of higher organisms: they contain "introns" and "exons" (no relation to gas stations or freeway ramps), and bacteria do not. Imagine you're reading a book and after thirty pages you come to pages of gibberish, scrambled letters that make no sense, then after a page or two the story continues; after a while, more gibberish, then the story goes on again. Exasperating for you, and exasperating as well for genetic engineers: reading along the DNA sequence of a gene, they suddenly encounter gibberish, then the gene goes on. These "nonsense" DNA regions are easy enough to recognize, because when the gene is translated into protein, the gibberish regions don't wind up reflected in the protein sequence.

Scientists call those gene regions that are actually expressed as protein *exons,* presumably for expressed regions, and the non-protein-coding gibberish sequences *introns,* or intervening regions.

Not only bacteria but many yeast genes lack introns. That is, all their DNA bases between the start and stop signals are expressed as protein.

Why do some genes have introns? No one knows. They are among the mysteries of molecular biology. How are they gotten rid of before translation? That is known. The introns are copied into the mRNA along with the expressed gene regions, then they are spliced out of the mRNA before it reaches the ribosomes, as shown in Figure 9 (next page), where expression is sketched for both bacteria and intron-containing organisms. The protein-coding DNA (i.e., the gene or exon) is denoted by thick lines, the non-coding DNA (i.e., introns) by thin lines, and hash marks show the gene start and stop signals. On the right side of the figure, E marks the exons, I the introns. Notice how much more compli-cated the higher organism's translation process is than the bac-terium's. The higher organisms's mRNA first is folded up inside particles called ribonucleoprotein particles so that the splice junc-tions (the filled circles in the figure) are aligned with each other. Enzymes now cut at the splice junctions, the introns are discarded —much as a film editor might cut certain scenes—and the remain-ing exons are spliced together. Net result: mRNA minus introns, after a lot of work that is still meaningless to us.

Whether or not we understand the existence of nonsense DNA, we are faced with it when we try to splice the gene of a higher organism into bacteria, which have no mechanism for removing the introns. If we shotgunned the genes as described in Chapter 4, we'd find gibberish expressed in our cloned bacterial colonies. The way around the problem is difficult: copy DNA.

We start by obtaining the mRNA from the cell *after* it has had the introns taken out, then make a DNA copy of it in a backward formation from the usual DNA-to-mRNA direction of synthesis carried out by RNA polymerase. As we mentioned before, nature supplies the enzyme for this backward formation, discovered by David Baltimore, the aptly named reverse transcriptase. The re-sulting DNA copy is called cDNA. The cDNA is then spliced into the bacterial plasmid, the plasmid is placed into *E. coli*, and we proceed as usual.

Reverse transcriptase makes it possible to do what otherwise would be impossible. Cloning some interferons or any large ani-mal-protein gene in bacteria could not have been done without it.

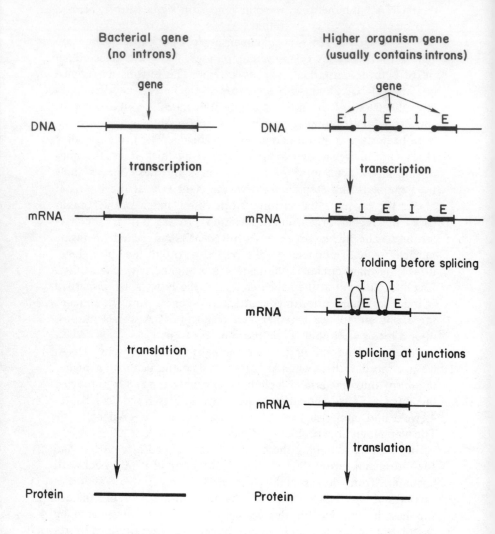

Figure 9. Comparison between the processing of bacterial mRNA and higher organism mRNA.

But it does not make these engineering feats easy; the procedure is far more tedious than it sounds. First, finding the mRNA for the desired protein is sometimes very difficult. If the protein is a rare one, then at any given moment very few if any of its mRNAs will be in the cell. If genetic engineers are very lucky, they find conditions under which the normally rare protein is made in abundance so the cell will contain a lot of the right mRNA. But the search for such conditions, as with the search for the mRNA needle in the cellular haystack, can take years. Making a cDNA copy from mRNA is also very tedious, as is preparing (or "tailing") the cDNA for insertion into *E. coli*. Right now, a project involving cDNA cloning—and that means many of those we've talked about—could take from one to several years or might not be feasible at all yet. This helps to explain why achieving expression of interferon was expected to take so long, why the early success was truly such an achievement and why we can't assume other feats will go as smoothly.

Improvements in procedure over the next few years should make cDNA cloning somewhat easier. Nevertheless, these uncertainties are serious ones for the research director, potential investor, or consumer wondering whether a new miracle product will soon be available.

SECONDARY METABOLITES:
Looking for the Fast Lane

Some of the greatest efforts and most revolutionary developments in genetic engineering's next phase will probably be in the improved genefacture of those vital products with unwieldy classification, "secondary metabolites" like alcohol and antibiotics, as well as primary metabolites like vitamins. But secondary metabolites, unlike many proteins, usually result from a long and quite delicately balanced series of reactions in the cell, so producing them presents special challenges. Biochemists call such progressive sequences of chemical reactions within a cell "metabolic pathways"; the movement on these pathways isn't necessarily from place to place, but from chemical to chemical in a chain of reactions.

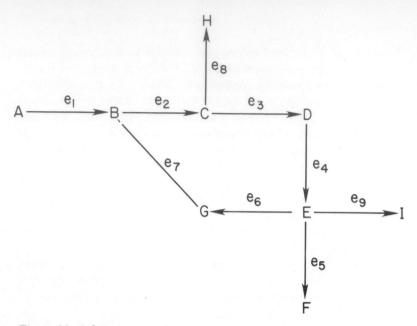

Figure 10. A fictitious metabolic pathway.

Figure 10 shows a fictional metabolic pathway, and letters A through I represent different chemicals produced in the reaction pathway. For example, if we were making alcohol in yeast, A would be glucose, the sugar/food source, and F would represent the secondary metabolite, alcohol. H and I might be chemical products necessary for the cell to function, and to make both of them requires all the intervening reactions: A to B to C to D to E to I, and the product H is made along the way. Remember that in the cell these reactions are catalyzed by enzymes, and this particular metabolic pathway involves nine of them—e_1 through e_9. Arrows show which way the reactions are moving.

If this sketch begins to look like the traffic pattern along city streets, these reactions behave much like flowing traffic, too. Think of A and B and the other capital letters as street intersections. Point F is the football stadium on the day of the big grudge match, and our problem is to get as many cars there just as quickly as possible. But there's always a bottleneck: the street between C and D has cars parked on both sides, so traffic narrows to one lane. That means that no matter how fast cars get from A to

B and from B to C and so on, they will arrive at the parking lot only at the slow crawl from which they get from C to D. This bottleneck principle actually applies in analyzing both traffic patterns and metabolic pathways.

Now look at Figure 10 as a series of chemicals and the reactions between them. The rates of reaction largely depend on which enzymes do the catalyzing: if e_3, which converts chemical C to chemical D, works at a slow crawl, then no matter how fast all the other enzymes along the pathway work to form the products B, C, E, F, getting to product F will take at least as much time as the slowest step, the formation of product D. Here is the bottleneck again, and having found it, we can see the direction in which solutions may lie.

The genetic engineer might find in another organism a better enzyme to change C to D and engineer it and its control sequence into the organism that makes our product. The new, fast-working enzyme will remove the bottleneck and the new rate of production will be on the scale of the *next* slowest reaction. Now we might get fancy, finding better enzymes for each step, putting strong promoters in front of their genes to get more of the new enzymes, and vastly increase the rate of production of F. This is precisely what genetic engineers are trying to do with alcohol production. Alcohol's metabolic pathways and enzymes are well-known, so we ought to be able to beef up enzyme concentrations in the slowest steps.

Another trick to get more product F: destroy the functions of enzymes e_8 and e_6, so that everything is funneled straight to F. Means to destroy enzyme function were developed long ago by randomly changing bases of their genes into different bases. This is the chemical meaning of mutation. Some altered base sequences might occur in the control region of the gene itself. Mutations in the gene might lead to an enzyme with a reduced rate of catalysis or none at all, meaning the cell either can't carry out that reaction or has to do it by some other pathway. In Figure 10, if by mutation we can stop enzyme e_8 from working, then everything that funnels into C must go on to D; none will be siphoned off into H. If we can also knock out e_6 and e_9 by mutation, everything that starts out as A must end up as the desired product, F.

Superficially this seems like the "best route" from A to F, but it may not be. Chemicals H and I might be necessary to the cell's continued life. Knock them out and we may kill a valuable colony, and to paraphrase the old detective-story line, dead cells don't metabolize. (If we wanted a legal definition of death for a single-cell organism, it might well be cessation of metabolism.) That means the genetic engineer is always walking a tightrope when tinkering with metabolic pathways in cells. Even if an alteration does not kill the cell, it might slow its growth, and that in turn could wipe out the benefits of speeding up the desired chemical reaction. Another disadvantage of this technique is that, when it is used, one must search through thousands of clone colonies to find those that have the right kind of mutations, and in trying to beef up production of other products, the right one might never be found—it might not even exist, due to the randomness of mutation.

On the success side of the ledger, the amino acid lysine has been produced in large amounts without unwanted side products using the random mutation technique.

Genetic engineering may revolutionize this old procedure, too. The scientist knows beforehand what enzymes are to be beefed up, and can therefore *direct* changes in the metabolic pathways by finding the right enzyme and engineering it into the desired organism. The genetic engineering of metabolic pathways adds a new bag of tricks to the production of important secondary metabolites. Eventually, whole new pathways or portions of these pathways might be engineered into an organism such as *E. coli*, which would then manufacture secondary metabolites foreign to it, just as the bacterium now manufactures foreign proteins. That could open up a whole new chemical genefacture industry.

FERMENTATION SCALE-UP:
The Easy Part Gets Hard

No matter how successful the laboratory wizard might be, his or her discoveries have no commercial importance unless they can be scaled-up to whatever industrial levels are required for a particular product. In the case of some drugs used in tiny quantities, scale-up may represent only a routine problem. In the case of large-volume products, that same scale-up may involve difficulties beyond any met in the lab.

In the laboratory it is usually easy to grow microorganisms in flasks of one or two liters, but the volumes of giant industrial fermenters often run to the thousands of liters. Why should that pose problems, as long as the growing colony is surrounded by nutrient? First, there's the problem of sterilization. It's easy to sterilize a laboratory flask—you put nutrient in it, stick it in a small autoclave (which kills off bacteria with heat), turn on the steam, wait half an hour, and it's done. How about a 10,000-gallon fermenter full of nutrient? Enough heat to do the job is hard to deliver, and even tiny leaks could pollute the nutrient broth with foreign organisms. One possible answer might be to develop "thermophilic" bacteria for large-scale jobs. They would thrive at a temperature that would kill intruding bacteria—unless, alas, the invaders also thrived at high temperature, at which most bacteria would die.

Further, genetically engineered microbes often grow more slowly than their "wild" counterparts (civilization always has its price), because they are using some energy that might otherwise go into reproduction to make a foreign protein. It doesn't take much slowing to have radical results, yielding a fermenter full of a wild invading strain instead of the one painstakingly engineered.

Suppose you inoculate a fermenter with a billion cells, of which 10 percent, or 100 million, are contaminant wild cells. Suppose the natural cells double every twenty minutes, but the redesigned cells take forty minutes. In 200 minutes—less than four hours—the fermenter will contain about 102 billion natural cells but only 29 billion engineered cells. The lesson: Even a tiny contamination by faster growing cells will invariably lead to crowding out of the slower growing ones. Even though the overwhelmed cells may still be making product, nutrient is expensive and most of it is going to nonproductive but very hungry contaminant cells.

A related problem could turn out to be even more serious. There is a natural tendency for organisms containing plasmids— as most genetically engineered bugs do—to throw off the plasmids eventually. That occurs as the cell doubles, when one daughter cell winds up without any copies of the plasmid. If the engineered features don't grant the cell better ability to survive, then the cells without plasmid, though initially few in number, may grow very much faster. Since plasmid-shedding is natural, it could become a chronic headache in genefacture.

In the laboratory there is a fairly easy solution to the problem, but it may not translate to industrial scale. Along with the desired gene, the genetic engineer splices resistance to a particular antibiotic into a plasmid. Cells then are grown in a medium containing the antibiotic. *Now* those that throw off the plasmid die. It's a neat solution on a small scale, but antibiotics are too expensive to be used in large fermenters.

A possible solution lies in the use of molecular switches. We mentioned one type earlier: the repressors. Repressors keep genes "turned off" until their proteins are needed. For example, if a cell normally metabolizes (eats) glucose but suddenly is in a medium where another sugar is available that it can eat, it will want to turn off the gene that codes for a glucose-metabolizing enzyme and turn on the gene coding for one that metabolizes the new sugar.

In industry, we might use molecular switches to keep the gene for our product turned off until the fermenter is filled to the brim with cells. Then the gene can be switched on and the cells will begin making our product all at once. The cells will be virtually all the engineered variety: no energy will have been lost during growth since the desired protein product is not made during the growing phase, and a few contaminants or a few cells tossing off plasmids will multiply out to a continued small percentage.

Some types of switches involve repressors, which are protein molecules, and a special region of the DNA called the operator region. The principle behind such operator-repressor switches can be seen under the schematic microscope of Figure 11. Figure 11a and 11b shows transcription as it was discussed in Chapter 4. An RNA polymerase molecule (solid square) binds to a promoter sequence P. The RNA polymerase then moves downstream and transcribes messenger RNA, shown in Figure 11b. But there is a new element here, between the promoter and the start of gene sequence G, a DNA sequence labeled O, the operator component of the molecular switch. (Some operator sequences actually exist inside the promoter region.) In Figure 11c we return to the transcription process before the mRNA is made, where the polymerase is bound to the promoter, as in Figure 11a. But here the operator has bound to it the protein molecule called a repressor, shown as an ellipse with a pie-shaped wedge cut out of it.

How does a repressor repress transcription of mRNA? By physically blocking the RNA polymerase from binding to the pro-

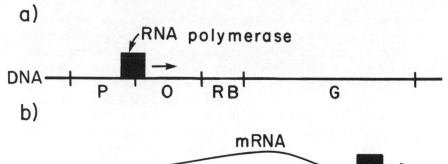

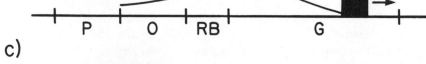

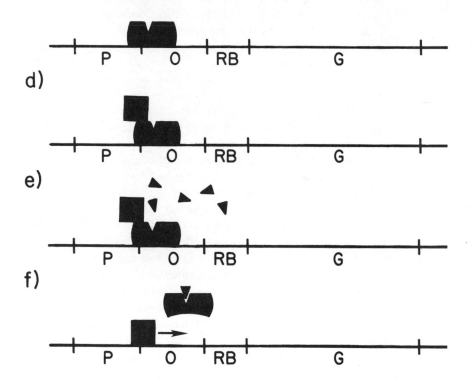

Figure 11. How repressor-operators act as molecular "switches" for transcription.

moter and/or by blocking the downstream movement of RNA polymerase, as shown in Figure 11d, thus halting protein production for that gene. If there is no sugar in the cell's medium that is metabolized by gene G's enzyme product, then there's no need to have a G switched on. Later the cell may find itself deep in that very sugar—shown in Figure 11e as little pie-wedges. Evolution has designed repressors so that they can bind to themselves "inducer" molecules, so named because they stimulate the production of enzymes governing their metabolism. Appropriately, the inducer in this case is the very sugar that is now to be metabolized. Once bound to the inducer, the repressor can no longer bind to the operator region of the DNA. In other words, the inducer binds, the repressor is released from the operator region, and RNA polymerase moves downstream from the promoter to transcribe the gene (Figure 11f) that will metabolize the sugar (inducer).

Molecular switches can be used in genetic engineering by splicing operator sites along with the repressor gene into the appropriate places on the plasmid containing the product gene. That will keep the gene repressed while the colony is growing up at the fastest rate possible. Once the colony is grown, the genetic engineer simply dumps inducer into the broth and transcription begins. Genetic engineers are now trying to develop operator-repressor switches that use inexpensive inducers.

Molecular switches may also prove useful in cases where some secondary metabolites produced in desired large quantities may poison the producing bacteria. In this case, we could grow a giant colony of microorganisms, then use the molecular switch to obtain a large amount of end-product by the time the colony died, whereupon we would begin growing a new colony. This would also enable us to overproduce end-products beyond the same range for the bacteria; it would not matter that they used up so much energy that they could not produce a viable amount of their own proteins.

Returning to fermentation scale-up, contamination is not the only problem that must be solved. It can also be difficult to get air spread uniformly throughout the medium in a large fermenter. *E. coli* and *B. subtilis* are aerobic bacteria, requiring oxygen in order to grow. Air might be bubbled into a giant fermenter containing the colony, but saturating the medium is difficult. Organisms such as yeast can grow with no oxygen (that is, anaerobically). Thus, genefacture of some products can be done in yeasts and other anaerobic organisms, where there are no aeration problems. It probably will take a few years to find the correct promoters and other expression-related features for most anaerobic organisms.

We have no doubt that these scale-up problems, serious as they are, are temporary. As genetic engineering becomes more sophisticated, it will provide techniques to make scale-up routine.

The Gene Age

What better place to end than where we began, but speaking in the new language—at least, no longer needing to rely on so many metaphors and analogies to explain where we stand and where we are going. *A gene is nothing more than a region along a DNA molecule.* Now we understand that it does not stand out from the non-gene regions because it is a different color, or because it is composed of different bases; it is distinguishable because the sequences of the same four bases are set in particular order; they *read* differently.

Seeing the DNA molecule as we first did in Chapter 2, as two chains whose links twine together in a spiral double helix with its linear array of bases, we notice that at one point we pass the sequence TATAAT. Ah! The Tata box. This must be a promoter region. Expect RNA polymerase to couple here and begin moving downstream, over the operator region where it will forge a messenger RNA's complementary initiation sequence and ribosome binding site. Passing the gene region, for example, we might notice the sequence TTT. When messenger RNA is transcribed along that sequence, the base uracil will replace the thymine (T) of the DNA, and that tips us off to what will happen way down on the ribosomes. "There's a gene coding for the amino acid phenylalanine," we remark, noting from the genetic code table that uracil (U) in the first position followed by U in the second must yield either phenylalanine or leucine. With U or C (cytosine) in the third position, we always get phenylalanine.

If we could jump over to the ribosomes, a protein chemist might be able to tell us from the order of the amino acids, in turn, what protein was being constructed there—maybe a short-chain hormone, maybe an enzyme such as glucoamylase, maybe even one of the enzymes that would eventually come back to the DNA to continue transcription—such as the enzymatic protein RNA polymerase.

But for now, just as friends on a long, arduous trip might entertain one another by talking about the local sights and formations and confirming the witness of their eyes with a road map, we can talk about landmarks we might pass in some gene regions.

MEDICINE

The first human protein synthesized was somatostatin, a hormone produced in the hypothalamus that affects release of other hormones in the pituitary that are crucial to human fertility. But it was chosen for the first synthesis because it is so short—only fourteen amino acids make up its chain, and we can read them off easily:

Ala-Gly-Cys-Lys-Asn-Phe-Phe-Trp-Lys-Thr-Phe-Thr-Ser-Cys-STOP-STOP. We know that the corresponding gene region will be forty-two (3 × 14) base pairs long, and that it will begin with one of the codes for alanine: CGA, CGG, CGT, or CGC. It will end in that double stop that could be coded by one of the termination signals in the code: ATT, ATC, or ACT.

Researchers *synthesized* that string of nucleotides, an easy procedure if the protein to be made is short. They then provided the synthetic gene with sticky ends using one of Boyer's terminal transferase enzymes, allowing them to stick (or splice) the gene into an *E. coli* plasmid. The final DNA segment to be spliced was fifty-two base pairs long: forty-two coding for somatostatin, and ten providing sticky ends and a little of the information needed for expression of the gene. Scientists spliced this creation into a plasmid that also had added the control region and much of the gene region from an enzyme called β-galactosidase. When the β-galactosidase promoter began functioning to produce that enzyme, it produced the human protein as a tail on the end of it. The somatostatin then was chemically cleaved from its β-galactosidase carrier to yield the completed human hormone.

But obviously that kind of method works only with a short chain, for tacking bases together is laborious. Human growth hormone is long by comparison—too long for synthesis of the gene. However, human growth hormone genes can be spliced into bacterial DNA. Figure 12 is a scanning electron micrograph of bacteria in their normal state. Figure 13, taken with a phase-contrast microscope, shows bacteria which have had human growth hormone genes inserted into their genetic array. The light white spots are the human growth hormone product which the bacteria have made along with their own normal proteins.

In practice, interferon was first synthesized by making a cDNA copy of the interferon mRNA from a strain of animal cells that overproduced interferon, thus making the usual needle-in-haystack search a little easier. The development of an mRNA overproducing strain and the screening for the proper clones were the difficult parts of this job.

With only a little new knowledge, it isn't hard to understand how these first, highly publicized genetic engineering feats were achieved.

Since those achievements a lot of progress has been made in finding strong promoters so many messenger copies would be made, and better ways for adjusting the promoters' distance from the desired gene to accomplish overproduction have been found. It all sounds similar to the tuning up of a complex piece of very precise machinery; it is.

Back in Chapter 1, we mentioned the possibility of creating completely safe vaccines against viruses. Here's how: Viruses contain only a small amount of DNA, typically on the order of 10,000 base pairs—the same order of magnitude as a bacterial plasmid's—split up into very few genes. Some of those genes code for manufacture of the virus's protective protein coat, on whose surface are those identifying markers that trigger antibody formation. On DNA molecules of this small size it is not difficult to find and clone the gene for one protein. (The 10,000 base pairs will be grouped into anywhere from five to twenty genes of somewhere around 1,000 base pairs each. Compare these 10,000 base pairs to one million base pairs in a bacterial DNA molecule or one billion in an animal's.)

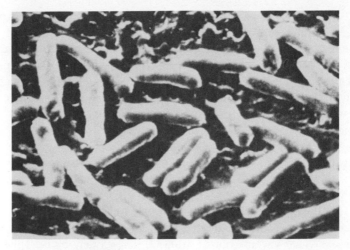

Figure 12. Normal bacteria.

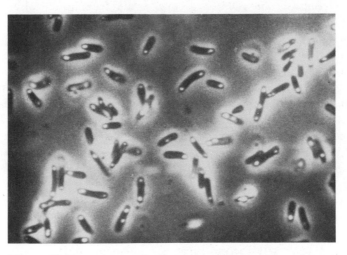

Figure 13. Bacteria genetically engineered to produce human growth hormone (white spots).

for
gene
hen
nce
n-
n

To make such a vaccine, we just chop up the DNA with restriction enzymes, shotgun the viral genes into *E. coli*, and find the protein-coat gene. How? Inject the virus into laboratory animals and isolate antibodies to the living virus. By means of a simple chemical reaction, the antibodies can be mixed with fluorescent or radioactive chemicals that will "tag" each antibody molecule. Now add the antibody carrying its warning beacon to each of the cloned *E. coli* colonies, and find the one the antibody sticks to—it's got the coat protein and the gene coding for it is in the bacterial plasmid. This rDNA-produced coat protein will now mimic a live- or killed-virus vaccine and will induce immune resistance to the whole live virus before infection.

The importance of such a recombinant DNA vaccine can be understood if you consider that if only one injection of the traditional killed vaccine out of 100,000 carries active virus, the results can be disastrous—not only to that one person per 100,000, but to the pharmaceutical companies as well, many of which have simply avoided making certain vaccines because of possible heavy lawsuits.

In contrast, the rDNA vaccines made from genetically engineered virus-coat proteins are 100 percent safe, since they contain only a small, harmless part of the whole virus, enough to marshal our immune systems against future viral attacks. There is a potential, in a fairly short period of time, for the development of brand-new vaccines against such viral infections as genital herpes, meningitis, and hepatitis, and related infections such as malaria. Further, we may radically cut the price on existing vaccines. Live or attenuated virus for vaccines must be grown in very expensive cultures. *E. coli*, remember, likes sugar water with a few additives.

On the cutting edge of such medical research are discoveries of possible someday "miracle" proteins such as longevin.[1] Described in a recent issue of the British journal *Nature*, longevin is a small protein consisting of two amino acid chain subunits, conveniently called the A and B chains. The possible miracle: this claim has been made falsely for milennia, but longevin appears to prolong life by slowing the aging process. Investigators, according to *Nature*, discovered that carp, noted for often living a century or more in a healthy state, release longevin's A-chain neuroendocrine cells in the gut. The B-chain? It is synthesized by *E. coli*, resident in the fish's internal organs! In some mice with identifiable genetic constitutions, the longevin did prolong life, the researchers claim in the article. But don't rush to the pharmacy yet: experiments so far have found that some injections of longevin into mice produce multiple tumors.

The major point here is that if longevin or some other protein proves a real inhibitor of the human aging process, the potential for manufacturing it or fine-tuning its composition through genetic engineering should now be obvious.

INDUSTRIAL CHEMICALS

We've dealt with these chemicals in some detail, but let's look over a couple in new detail, beginning with glucose isomerase, the enzyme that converts the less-sweet sugar found in grapes and blood to the supersweet fructose, which is used to make high fructose corn syrup. Genefacturing this should be a quick process, although there is bound to be competition for that reason among the many new rDNA firms.

We begin by finding the enzyme in a microorganism; since it occurs naturally at this level of life, the gene coding for it will contain no introns, those gibberish sequences that can make producing animal proteins so difficult.

Now we select as host a mutant microorganism that *does not* produce glucose isomerase; that will be important when we try to find the gene, because we're going to do this job by shotgunning. In other words, we take an organism that *does* produce glucose isomerase and take out its DNA, cut up the DNA with restriction enzymes into many little gene-sized pieces (whose identities we don't yet know), splice all the pieces into plasmids, and then shotgun all these plasmid vectors into a colony of host microorganisms that *don't* produce this enzyme.

Next we follow the biologist's long-known cloning process: Dilute the colony in solution so thinly that on the average each cell will have a centimeter of space between it and other cells, and grow up separate daughter-cell colonies from each original. One colony will be producing glucose isomerase (if all has gone right), for its original parent will have taken up the gene-bearing plasmid. This is exactly the procedure we studied under the schematic microscope as shotgunning. There is a simple color assay test to find which colony has the desired plasmid—a chemical combination that becomes colored when mixed with glucose isomerase.

The job's not done yet. We have plenty of genes available that code for glucose isomerase, but it's not likely they're tuned up just right. This could be where we separate winners from losers: everyone will be trying to find strong promoters and to adjust promoter-to-gene distance for overproduction. Some experimenters will be trying to produce the enzyme in *Bacillus subtilis* or another organism, so that the glucose isomerase can be secreted out of the producing colony and save the time and expense of purification. A few companies now are studying the secretion systems of *B. subtilis*, and within one to three years there should be generalized means to secrete protein from these cells.

Now let's consider a very different kind of protein, both in type and in the uses to which it is already put: rennin. Glucose isomerase is fairly easy to find; rennin is harder. Its traditional sources are the stomachs of young calves and other animals, where it is a natural enzyme. Glucose isomerase is a bacterial enzyme, rennin an animal one—which means it probably has introns and exons interrupting its gene sequences, barring easy production in bacterial factories.

What is its value? Rennin is the enzyme used to make some of the world's finest cheeses, ones that have been made in the same way for hundreds of years. Substitute enzymes just don't do as well. Europeans eat a lot more veal than Americans, so they have always had a larger supply of rennin, even though their stock is also unstable and in short supply. Americans have made do with substitute cheeses, imported cheeses, or have paid the extremely high price for rennin. How do we tackle this problem?

The protein is a large one, so we can't synthesize the gene sequence coding for it by stringing the right base pairs together, and there are probably introns. That means we have to go to cDNA cloning. This way it might take a year or two to get a cDNA copy of the gene from the mRNA instead of the one to two months for a gene that can be obtained by shotgunning. Once the cDNA gene copy is obtained, we shotgun it into *E. coli* or *B. subtilis*. Then we get strong promoters, and adjust the promoter-to-gene distance for overproduction. If we succeed, we will not only have an inexpensive enzyme whose current high price is a major factor in cheese production costs, but we will have a plentiful supply of an enzyme now quite rare—not the miracle of interferon, perhaps, but in its own way a real "sports car" made while *E. coli* is just trying to go about its business of subdividing—shades of Frank Lloyd Levitt.

Hypothetical? The following is a portion of an Associated Press story carried by *The New York Times*[2] in 1982:

LEXINGTON, Mass., May 4 (AP)—Because of a new commercial enzyme, better-quality cheese may soon find its way to American tables, according to Dr. Orrie Friedman, president of a Lexington-based biotechnology company.

Dr. Friedman, head of Collaborative Research, Inc., said his concern had cloned the gene for rennin, a natural enzyme used to curdle milk in the cheese-making process. . . .

"Rennin production is expected to be one of the first commercial applications of gene-splitting technology," he said. "Until now, most of the new products made through genetic engineering have been of medical importance, such as hormones and vaccines.

"The cloning of rennin using bacteria and yeast [used because it has a better chance of FDA approval as an organism for food products] was done by making a copy of the gene that codes for rennin in the calf's stomach," he explained. "The copy of the gene was spliced into a 'vector' such as a ring-shaped molecule called a plasmid. The vectors were then used in the host cells to convert them into rennin-producing cells."

Daily, such stories are converting speculation on hypotheticals into hard news, and many of the prospects we've discussed will undoubtedly no longer be hypothetical by the time this book is out. In those cases, we hope at least to offer some better understanding of the news.

In Table 3 (next page), we offer our own assessment of the uses of many industrial and medical enzymes and other proteins in terms of when they might be available.

Table 3. A Forecast of Genetically Engineered Products

Name	Source of Gene	Commercial Uses	Time to Carry Out Necessary Genetic Engineering*
I. Enzymes			
α-amylase and other amylases	Bacteria; fungi; thermophilic bacteria	Degrade starches into glucose, which is used as a carbon source nutrient for microorganism growth. Examples: yeast for alcohol production, other organisms for single-cell protein production for human nutrition	1–3 years
Cellulases	Fungi; micro-organisms in termite and cow gut	Degrade cellulose in wood, waste paper into glucose, which is used as a carbon-source nutrient for microorganism growth	3–6 years
Glucose isomerase	Various species of bacteria	Convert glucose into the popular artificial sweetener fructose	1–3 years
Proteases	Various species of bacteria	Enzyme detergents; digestive aids; tanning of leather	1–3 years
Rennin	Cells of calf intestine	Cheese manufacture	1–3 years

Strepto-kinase	Bacteria	Treatment of heart conditions and embolisms, blood clots	1–3 years
Urokinase	Human cells	Treatment of blood clots, embolisms, possibly heart conditions	2–5 years
Lactase	Bacteria; animal sources	An addition to powdered cow's milk to make it digestible for the large numbers of humans who cannot drink cow's milk	1–5 years

II. Nonenzyme Proteins

Interferons	Human cells	There are many types of interferons in human cells. Separately or in combination, they may be found to be powerful cures for viral diseases. They may also find uses in treatment of various cancers	2–8 years
Insulin	Human cells	Control of diabetes	1–4 years
Virus coat proteins	Various disease-causing viruses	Manufacture of inexpensive vaccines against animal and human viral diseases	1–3 years

* Some of these products have already been produced commercially, either by traditional means or, in a few cases, by recombinant DNA.

Table 3 (*continued*)

Name	Source of Gene	Commercial Uses	Time to Carry Out Necessary Genetic Engineering*
II. Nonenzyme Proteins (*continued*)			
Human growth hormone	Human cells	Treatment of dwarfism	1–3 years
Epidermal growth factor	Human or animal cells	Research use in animal-cell growth media; possible addition to synthetic infant formulas to make them more similar to mother's milk; treatment of ulcers	1–3 years
Serum albumin	Human cells	Treatment of shock	1–4 years
Cytokines	Human cells	Possible use in treatment of immune disorders	2–8 years
17 proteins involved in nitrogen fixation (some are enzymes)	The *rhizobium* bacterium; other bacteria	These proteins to be engineered in plants to reduce or eliminate the need for expensive manufactured fertilizers	5–10 years

III. Primary, Secondary Metabolites, and Other Nonprotein Cellular Products

Product	Source	Use	Time
Ethanol (alcohol)	Yeast, *clostridium thermocellum*, or other thermophiles	Industrial and automotive fuel; starting chemical (chemical feedstock) for some industrial processes	1–6 years
Acetone	*Clostridium acetobutylicum*	Industrial solvent; used as chemical feedstock	1–6 years
Glycerol	Yeast; algae	Chemical feedstock	1–6 years
Alkene oxides	Possibly fungi or bacteria	Plastics manufacture, process involves enzymes found in many microorganisms	3–6 years
Methane	Methanobacteria	Automotive, household, or industrial fuel; chemical feedstock	3–6 years
Methanol		Chemical feedstock for plastics and chemical industry	3–6 years
Citric acid		Flavoring agent in soft drinks and other foods	1–4 years
Amino acids	Various microorganisms	Nutritional additives to animal feeds and human food	1–6 years
Antibiotics	Various microorganisms	Cure of bacterial infections	2–6 years
Steroids (cortisone; hydrocortisone)	Animal cells	Reduce inflammation caused by injuries and diseases	2–6 years

AGRICULTURE

Of the twenty amino acids, ten are considered essential to human nutrition. Because we cannot manufacture them ourselves, they must come from the food we eat. The essentials for humans are lysine (first mentioned in Chapter 1, because corn is deficient in it), isoleucine, leucine, methionine, phenylalanine, threonine, histidine, tryptophan, arginine, and valine.

Let's consider two ways of getting more of these to feed the world: beefing up food crops now deficient in them to produce more, and directly producing them in large quantity.

Such worldwide staples as rice, wheat, sorghum, and beans are all deficient in at least one of the essential amino acids—generally lysine and tryptophan. When our genes begin calling for production of vital proteins, if a single amino acid is missing, production at the ribosome is stalled so the protein cannot be made. It takes only days for such nutritional deficiences to be felt. But all the foods just mentioned have one thing in common: they are seeds. In addition to the plant's genetic information, they contain large quantities of "storage proteins." Normally, the nonseed parts of plants contain little protein. In terms of the plant's growth, the storage proteins are used to supply the seed sprout with the amino acids *it* needs to grow until it breaks out of the soil and can begin making amino acids through the energy derived from photosynthesis and the nutritional elements found in soil (such as fixed nitrogen). It is these abundant storage proteins we are really after when we eat corn, rice, or wheat, whose kernels are the seeds containing these proteins.

The fact that many plants can be grown from a single cell makes our job vastly easier, because if we can alter the amino acid balance in just one cell, we should be able to grow a new plant variety from it. How to alter that balance? Conceptually, that is very simple. Back in Chapter 1 we mentioned that genetic illnesses could be caused by changing just one of the three DNA triplets coding from a particular amino acid, because then the wrong amino acid would be hooked in place. That turns out to be the secret of genetically engineering seed crops as well. Looking back at the genetic code, you'll notice that a great variety of triplets will code for arginine, which is overabundant in most crop plants and is similar in structure to lysine, which is vital. Among those triplets are two very similar to those coding for lysine: AGA or AGG coding for arginine would become AAA or AAG, yielding lysine, if just one G or C were replaced by an A at the appropriate site on one of the plant's gene regions.

Biochemical techniques already exist to make these base replacements. All we need to do is remove the gene's coding for the most abundant storage proteins and replace the key base in the code for arginine by known techniques by using *E. coli* as a host for the operation. We then remove the storage-protein genes containing the new lysine-coding sequences from *E. coli* and splice them back into the plant cells growing in culture along with an appropriate promoter for the storage protein. We should not be many years from being able to do this on a scale sufficient to increase sharply the nutritional content of many of these plants.

But let's consider the direct manufacture of these essential amino acids on a large scale. We can already sketch out what will be necessary, using lysine as an example. Lysine is one of those metabolites whose production causes a reaction that inhibits its own further production. When a cell produces lysine, the lysine normally inhibits the action of an enzyme called aspartate kinase, which is vital to the enzymatic pathway that yields lysine. That makes lysine what is known as *feedback inhibited.*

What follows should sound familiar. It follows directly from our discussion of enzymatic pathways in Chapter 9, and it might be a good place to compare our strategy for increasing lysine production to that discussed for altering any metabolic pathway.

During cellular metabolism, the particular pathway that we're concerned with here starts with aspartic acid. The enzyme aspartate kinase breaks down aspartic acid into a series of related compounds, one of which comes under the action of yet another enzyme, homoserine dehydrogenase, which, in turn, degrades the compound into lysine and homoserine. The homoserine then breaks down into methionine and threonine. Thus, three amino acids are produced along this metabolic pathway, and large concentrations of both threonine and lysine inhibit further amino acid production. This is how the cell stops itself from making too much of some amino acids.

Recalling our earlier discussion of enzyme action along metabolic pathways, the analogy will be seen in what follows. We can get more lysine in one of two ways: first, find a mutation in the gene coding for homoserine dehydrogenase that yields an inactive enzyme. Then, the *only* amino acid that will exit on this pathway is lysine. Secondly, we might mutate the gene coding for aspartate kinase, the vital enzyme that set the whole action in motion, so that it will continue to act no matter how much lysine is present. Or we can do both: mutate genes for homoserine dehydrogenase so the pathway will yield only lysine, and mutate the gene coding for aspartate kinase so that it will do its job no matter how much lysine is present; that is, we'll remove the feedback inhibition.

Certainly, this last discussion has taken us to the technical limit we can reach in this book, but in simpler terms, consider these other quite lively possibilities for genetic engineering.

Nitrogen fixation is certainly the single most important long-term goal of agrigenetic engineering, but it should also be possible to clone into plant cells the genes offering protection against many devastating crop diseases, or genes that will yield proteins fatal to crop-eating insects, or proteins that will simply make the plant unappealing to the pests. If protection against herbicides can be spliced into the gene complement of food plants, then herbicides harmless to crops but fatal to weeds could be sprayed on farmland.

We also discussed human growth hormone at some length. Varieties of growth hormones for livestock are now being investigated, and a few (such as bovine growth hormone) will soon be available. Others that will speed the growth of food animals should be available within three years. Most of these growth hormones are fairly short molecules, so figuring out their corresponding gene sequence should not be difficult. If the hormones do not need post-transcriptional modifications, they can be grown in *E. coli*. If they do, their genes may be spliced into yeast or animal cells. The purified hormone would then be injected into animals.

THE CUTTING EDGE:
The War Against Cancer

Generally we've discussed the commercial applications of recombinant DNA, and we think that understanding the complex concepts involved in all genetic engineering is made somewhat easier by looking at the relatively simple creatures used in industrial genefacture. However, recombinant DNA technology has had profound implications for basic research in molecular biology.

Some people think of "basic research" as referring to the arcane and incomprehensible, the work scientists do in their search for a "truth" whose meaning is hard for the rest of us to discern. Mention cancer, on the other hand, and everyone's interest quickens—yet research into the causes of cancer is just such basic research, and an area in which recombinant DNA techniques have brought scientists to major discoveries. In fact, recent achievements using these techniques have led some scientists to predict that 1984 will be the year the cause of cancer is discovered.

In brief, here are some of the important findings about cancer that could not have been made without recombinant DNA.

Certain animal viruses cause cancer—that much has been known for years. When these viruses infect an animal cell, they can cause it to divide and grow wildly. Such uncontrolled growth is the major distinguishing characteristic of cancer cells, threatening death to the entire organism from tumors or other factors associated with the abnormally growing cells.

But it has also been known for years that cancers can occur in the absence of viruses. Using recombinant DNA and other modern techniques, the causes of some viral and non-viral cancers have been traced to a single source—cancer genes, often called oncogenes. By such methods as DNA sequencing, the genes of at least one cancer virus have been studied extensively.

Harvey's sarcoma, a rodent cancer virus, has become something of a laboratory celebrity because of its importance in the hunt for the cause of cancer. A virus, remember, is a relatively simple structure: a short piece of DNA or RNA, usually protected by a protein coat. Its genetic instructions are few: some of its genes code for the protein coat and some instructions are directed toward taking over the host's protein-making machinery so that it can be used for the virus's reproduction. Scientists made a remarkable discovery about Harvey's sarcoma: the virus contains an animal gene, and that same gene was one found in the DNA of normal, non-cancerous animals—including humans. Further, the gene was identified as involved in regulating animal cell growth. In effect, this viral DNA was a naturally occurring example of recombined DNA. If invading viruses normally insert their DNA instructions into the host, this is an example of the opposite—the virus contains pieces of the host's genetic information. Why is it there? We still don't know, but this is the kind of discovery that makes a scientist's imagination soar. Cancer is abnormal cell growth. Suddenly we find that a cancer-causing virus contains a normal animal gene whose protein product is involved in *regulating* cell growth. Consider this scenario: perhaps a particular virus infects a human cell and takes over its protein-making machinery; now the animal growth-regulating protein from the virus is overproduced, causing the cell to grow out of control. It has been shown that cancers caused by Harvey's sarcoma virus indeed are associated with high levels of growth-regulating protein.

Other cancer viruses were examined and they, too, possess genes from normal human and other animal cells. We now presume that these oncogenes are also involved in the regulation of cell growth, though that still must be proven.

More speculative evidence also points to overproduction of normal animal-cell proteins as a cause of cancer. Long before oncogenes were found, researchers knew that chromosomal mutation was involved in the development of at least one form of human cancer, lymphoma. Many lymphoma patients had a chromosomal mutation called a translocation, which involves the movement of very large regions of DNA, including perhaps hundreds of genes, to a new location along the chromosomal DNA.[3] In this case, the amount of DNA moved is so large that the translocation can be seen in a microscope. That made its existence known well before the development of such exquisitely subtle techniques as direct DNA sequencing, now used to detect and study mutations.

Initially, such a translocation's association with cancer might seem wholly unrelated to oncogenes. But now investigators at both Harvard and Philadelphia's Wistar Institute have discovered that the translocation occurs at the very point where the existence of an oncogene had already been postulated. More interestingly, the translocation placed this postulated oncogene right next to a *strong promoter*, used in the normal cell to produce antibodies in large amounts. Thus, the explanation for lymphoma cancer also seems to tie in to overproduction of a protein involved in growth regulation; it is not yet proven that the postulated oncogene in fact regulates growth.

The oncogenes responsible for Harvey's sarcoma and some other oncogenes show another interesting property. Though nearly identical to normal animal genes, they contain a few mutations, base changes that code for a different amino acid in the resulting protein. Researchers have found that a single mutation changing a glycine amino acid into other amino acids makes an oncogene more active. The mutated oncogene need not be overproduced to cause cancer but can do so in very small quantities.

Recombinant DNA has become the most powerful tool used in the study of oncogenes, their DNA sequences, mutations, and relationships to one another. Typically, genes are present in an animal cell in at most a few copies. An average gene is about a thousand bases long. The whole human chromosome is more than a billion bases long. To study a single oncogene amid all the other DNA is almost impossible. Imagine that a researcher needs DNA weighing only a milligram—a grain of salt's worth—to carry out experiments on a given gene. Deriving it would require 1,000 grams of DNA, and that would be the entire amount of DNA found in a 220-pound human. Consider that cell cultures are grown in little round petri dishes (with a diameter of four inches, on which cells can grow only one-eighth of an inch deep) and the enormity of the problem becomes obvious.

Now the researcher takes a milligram or less of DNA and shotguns it into *E. coli* by methods we've already discussed. Next, the gene is located, and the clone containing the desired gene is grown up in any quantity desired and easily spliced out of the plasmid DNA on which it had been cloned. Without this incredible *amplification* provided by recombinant DNA techniques, these discoveries about oncogenes could not have been made.

Evidence is mounting rapidly that two closely related events— overproduction of normal cell growth-regulating proteins and mutation of smaller amounts of growth-regulating proteins—are the causes of cancer. Within the next year the lines of evidence could converge to prove it.

Causes are not cures. The cause of cancer may soon be found, but we must not be overoptimistic about finding the cure. Nevertheless, not long ago it was feared that mutations of virtually any gene might lead to cancer. That might make isolating a cause and finding a cure impossible. This evidence suggests that only a limited number of events involved in growth regulation are responsible for tumor formation, and that is exciting because it offers hope that a cure for our most feared scourge *will* be found. Such a discovery would bring to a close one of the major quests in the history of science.

CONCLUSION

If there is one truth that emerges in this book, repeating over and over like a base theme in endless variation, it is the universality of this language that all life on Earth holds in common, the language we've been describing, translating, trying to speak. And that suggests as well the great uniformity, congruity, or at least harmony to the notion of life that underlies the equally awesome differences between individual humans or between humans and other creatures, complex or microbial.

Why can humans metabolize the insulin of pigs and cows? Because the DNA chains of those homely animals along that particular gene-region are nearly identical to our own. But there is more similarity between all of "us" than that. An eye is an eye and a hair is a hair: there is more alike in the DNA sequences that code for these complex creations in different animals than there is dissimilar. The greater dissimilarity is between the eye and hair: how do identical DNA molecules "know" to order eye-building proteins at one place and hair-building proteins at another? The incredible process of cell differentiation, alluded to by several scientists in this book, still is one of the great unsolved biological mysteries. But not forever, not now that we've learned to speak the language, however crudely at this stage. At the level we've been traveling, of course, our *bodies* know the language fluently. Our "minds" may not, but our brain cells do. At that level, "we" speak the same language as the virus. If we did not, the virus could not infect us by taking over our own cells' mechanisms with such well-understood instructions. But then, we could also not have turned the tables, finally figuring out this great key to a boundless source of knowledge, energy, and tools that will change us forever.

Notes and Further Reading

Chapter 1: Industry Comes to Life

[1] *Nature*, December 1982. Copyright Macmillan Journals Ltd.

[2] *Impacts of Applied Genetics*, referred to throughout Chapter 1, was published by Congress's Office of Technology Assessment in 1981 and is for sale by the Superintendent of Documents, U.S. Government Printing Office, Washington, DC 20402.

[3] *Chemical Week*, Feb. 9, 1983.

[4] *Scientific American*, special issue on "Industrial Microbiology," September 1981. Reprinted with permission. This issue contains perhaps the most comprehensive look at industrial uses of genetic engineering so far.

[5] Values of products not otherwise attributed in this book are from the U.S. Department of Commerce's *Annual Survey of Manufactures*.

[6] "Cellulases: Biosynthesis and Applications," by Dewey D. Y. Ryu and Mary Mandels, *Enzyme and Microbial Technology*, Vol. II (1980), pp. 91–102, contains Natick's alcohol production estimates. Reprinted by permission of the publishers, Butter & Co. (Publishers) Ltd. ©.

[7] "Biomass as a Source of Chemical Feedstocks: An Economic Evaluation," by B. O. Palsson, S. Fathi-Afshar, D. F. Rudd, and E. N. Lightfood, *Science*, 31 July 1981, pp. 513–17. Copyright 1981 by the American Association for the Advancement of Science. Reprinted with permission.

[8] Alvin Toffler, *Future Shock* (New York: Random House, 1970).

[9] *Scientific American, op. cit.*

[10] Dr. Thomas E. Wagner has made his comments on the "unnatural" qualities bred by ordinary means into dairy cattle on several occasions, here quoted from a United Press International story of Dec. 10, 1982. Reprinted with permission.

[11] "DNA Science Inc.: First Casualty in the Biotechnology Derby," by Colin Norman, *Science*, 4 September 1981, pp. 1087, 1088, 1090. Copyright 1981 by the American Association for the Advancement of Science. Reprinted with permission.

Chapter 3: Genes and Genies

[1] *The Double Helix*, by James Watson, Atheneum, New York, 1968, is his much-acclaimed memoir of the race to discover DNA's structure. Reprinted with permission. For readers with a background in biochemistry, *Molecular Biology of the Gene*, by James Watson, 3rd ed., W. A. Benjamin (Menlo Park, Calif., 1976) is considered the classic textbook on the subject.

[2] *The Eighth Day of Creation: Makers of the Revolution in Biology*, by Horace Freeland Judson (New York: Simon and Schuster, 1979), is an excellent and readable chronicle of the recent history of genetics, for those interested in an in-depth account of the people and events involved. Copyright © 1979 by Horace Freedland Judson. Reprinted by permission of Simon & Schuster, a division of Gulf & Western Corporation.

[3] *Double Helix, op. cit.*

[4] *Eighth Day of Creation, op. cit.*, p. 264.

[5] *The Wall Street Journal*, May 20, 1982.

[6] John Hunt, *Generations of Men* (Atlantic, Little, Brown, 1956).

Chapter 5: Bull, Bear, and Clone

[1] *The Wall Street Journal*, "Ex-Convict and Microbiologists Join Forces to Ride the Genetic-Engineering Wave," March 4, 1981.

[2] *Time* magazine cover story, March 9, 1981.

Chapter 6: Chimeras

[1] The American Patent Law Association's *amicus curiae* brief is filed in the Supreme Court of the United States, October term, 1979, No. 79–136, in the case of *Sidney A. Diamond, Commissioner of*

Patents and Trademarks v. *Ananda M. Chakrabarty.* American Patent Law Association, 2001 Jefferson Davis Highway, Arlington, Va. 22202. Reprinted with permission.

[2] *Miracle Drug: The Inner History of Penicillin,* by David Masters (London: Ayer and Spotswood, 1946).

Chapter 7: *Academe, Inc.*

[1] "Speeding the Transfer of Technology: Business, Science and the Universities" (occasional paper), Derek Bok and Donald Kennedy, Committee for Corporate Support of Private Universities, Inc., 125 High St., Boston, MA 02110. (Copy provided by Dr. Kennedy.) Reprinted with permission.

[2] Dr. Kennedy's congressional testimony was given to the U.S. House of Representatives Committee on Science and Technology Subcommittee on Investigations and Oversight, June 8, 1981. Kennedy, who provided the transcript, represented Stanford, the Association of American Universities, and the National Association of State Universities and Land-Grant Colleges.

[3] Joseph Kraft, Field Newspapers, 1981.

Chapter 8: *Dangerous Acquaintances*

[1] Ray Thornton's statements on risks were made to the September 10–11, 1981, meeting of the National Institutes of Health Recombinant Advisory Committee. (Copy provided by NIH.)

[2] Laboratory-Associated Infections: Summary and Analysis of 3,921 Cases," by Robert M. Pike, *Health Laboratory Science,* Vol. 13, No. 2 (April 1976), pp. 105–14. Also, Laboratory-Associated Infections: Incidence, Fatalities, Causes, and Prevention, Robert M. Pike, *Annual Review of Microbiology,* 1979.

[3] "The Detrick Experience as a Guide to the Probable Efficacy of P4 Microbiological Containment Facilities for Studies on Microbial Recombinant DNA Molecules," A. G. Wedum, M.D., concluding research sponsored by the National Cancer Institute. (Manuscript of Jan. 20, 1976, provided by NIH.)

[4] *Impacts of Applied Genetics, op. cit.,* cited in Chapter 1 notes.

[5] "Ordinance for the Use of Recombinant DNA Technology in the City of Cambridge," No. 955, passed April 27, 1981.

[6] "The Specter of Eugenics," by Charles Frankel, was reprinted from *Commentary,* March 1974, by permission; all rights reserved.

Chapter 10: The Gene Age

[1] "The Elixir of Life" contains information on longevin. Reprinted by permission from *Nature*, Vol. 296, No. 5856, April 1, 1982. Copyright © 1982 Macmillan Journals Limited.

[2] *The New York Times* carried the Associated Press story on rennin on May 5, 1982. © 1982 by the New York Times Company. Reprinted with permission by *The New York Times* and the Associated Press.

[3] "Translocation and Rearrangements of the c-*myc* Oncogene Locus in Human Undifferentiated B-Cell Lymphomas," *Science*, Feb. 25, 1983. Copyright 1983 by the American Association for the Advancement of Science. Reprinted with permission.

Index